SONORA
Its Geographical Personality

SONORA
Its Geographical Personality

ROBERT C. WEST

 University of Texas Press, Austin

Library of Congress Cataloging-in-Publication Data

West, Robert Cooper, 1913–
    Sonora : its geographical personality / Robert C. West.—1st ed.
      p. cm.
    Includes bibliographical references and index.
    ISBN 0-292-76538-X
    1. Sonora (Mexico : State)—Historical geography. 2. Sonora
(Mexico : State)—History. 3. Mexico—History—Colonial period,
1540–1810. 4. Human geography—Mexico—Sonora (State) 5. Mineral
industries—Mexico—Sonora (State)—History. 6. Agriculture—
Economic aspects—Mexico—Sonora (State) I. Title.
F1346.W47 1993
972'.17—dc20   92-12490
                                        CIP

Contents

Tables

Abbreviations Used

AGI   Archivo General de Indias, Seville

AF-BNM   Archivo Franciscano, Biblioteca Nacional, Mexico City

AGN   Archivo General de la Nación, Mexico City

AHH   Archivo Histórico de Hacienda, Mexico City

AMP   Archivo Municipal del Parral, Chihuahua, Mexico

BNM-MS   Biblioteca Nacional de Madrid, MS Div.

Preface

The idea for this book developed while I was a visiting professor at the University of Arizona, Tucson, in 1972 and again in 1976. Given my training and academic interest in Latin American cultures, neighboring Sonora naturally beckoned. My familiarity with that part of northwestern Mexico stems mainly from the many field excursions into the state with students and faculty from Arizona and other universities over a period of nearly twenty years. I am grateful for the stimulating company of those field companions, especially for their keen observations and perceptive knowledge of the history and geography of that fascinating part of Mexico. Equally important was the encouragement of members of the Centro Nacional del Noroeste, a branch of the Instituto Nacional de Antropología e Historia, based in Hermosillo.

My introduction to the colonial records of Sonora came through Charles W. Polzer, S.J., organizer and present director of the Documentary Relations of the Southwest project at the University of Arizona. Father Polzer permitted me to peruse the large quantity of microfilm of colonial church and secular documents currently housed in the Archivo Histórico de Hacienda, Mexico City. Also in Tucson, in the University of Arizona Library Special Collections, I had access to microfilm of colonial documents on Sonora from the Archivo Municipal del Parral, Chihuahua, where I had worked many years on another topic. Microfilm and transcripts of documents in the Bancroft Library, Berkeley, California, and in the Benson Latin American Collection, University of Texas, Austin, were also consulted.

Subsequent archival research led me to the Archivo General de la Nación and the Archivo Franciscano (in the Biblioteca Nacional), both in

Mexico City. Finally, the ultimate goal of any Latin Americanist historian or historical geographer was reached when I visited the Archivo General de Indias, Seville. I spent several weeks there reading colonial documents pertaining to Sonora from various *ramos*, or sections, of that great collection.

Published materials on Sonora are abundant and varied; only a small number of them could be cited in the text and listed in the accompanying bibliography. I gathered substantial information on colonial Sonora and northern Mexico from the many writings of the Spanish historian Luis Navarro García, who has ready access to the Archivo General de Indias in Seville. Among North American scholars the Jesuit missions of northwestern Mexico have been a favored topic for research, but comparatively little has been written on colonial mining in Sonora, a subject central to the present study.

I use historical geography in this book to describe the character, or "personality," of Sonora and to illustrate how the cultural landscape of a given region in northwestern Mexico has developed through time, especially during the Spanish colonial period; but I bring that development up to the present day. My method is traditional and genetic, closely allied to Clifford Darby's vertical approach to landscape evolution.[1]

The term "personality" as applied to the physical and cultural character of a given geographical area was popularized nearly a century ago by French regional geographer Paul Vidal de la Blache, in his *Tableau de la géographie de la France*.[2] Gary Dunbar has traced the usage of the term by various geographers, mainly U.S. and British, and cites as examples Carl Sauer's seminal article, "The Personality of Mexico" (1941), and E. Estyn Evans's *The Personality of Ireland* (1973, 1981), among others.[3] In essence, the "geographical personality" of an area derives from the *genre de vie*, or way of life, of its inhabitants as adapted to the physical characteristics of the land. Again, Paul Vidal de la Blache was one of the first to apply the term *genre de vie* in geographical study;[4] the term was further defined by another French geographer, Max Sorre, who expanded its meaning to include not only the rural scene, but also an industrial and urban one.[5]

In tracing the history of Sonora's human development, I came to realize early on that today the state presents a dual geographical personality in terms of area: (1) an eastern subhumid, mountainous section incised by narrow river valleys, inhabited aboriginally by native farmers, in the seventeenth and eighteenth centuries occupied by Jesuit missionaries and Spanish miners-stockmen, and today living largely in its colonial past; and

(2) a western desert, most of which was sparsely populated by aboriginal nomads, exploited in the late eighteenth century by Spanish and Indian gold seekers, but today characterized in places by recently developed, government-sponsored modern irrigated agriculture, which has given rise to dense farming population, urbanization, and industrialization.

For technical assistance in preparing the manuscript I am indebted to Clifford P. Duplechin, senior cartographer, Mary Lee Eggart, artist/ research associate, and Maudrie Eldrige and Adriane Kramer, word processors.

SONORA
Its Geographical Personality

CHAPTER 1

Physical and Biotic Aspects of Sonora

Physically Sonora is composed of two distinct areas: to the west an arid zone of low, scattered mountains separated by extensive plains; and to the east a semiarid to subhumid mountainous region that flanks the western edge of the Sierra Madre Occidental (fig. 1). Each of these areas has undergone different cultural developments.

Eastern Sonora, or La Serrana

Composed of a series of north-south-trending ranges separated by narrow valleys, eastern Sonora, often called La Serrana, until recently was the more important of the two areas in terms of human occupation and economy. The river valleys, filled with narrow strips of Recent alluvium bordered by Tertiary and Pleistocene gravel terraces, have been the sites of agriculture and permanent settlement since long before the Spanish period, and during colonial days they were the main producers of food in Sonora. The adjacent mountain ranges are rugged, rising as up-faulted blocks eight hundred to one thousand meters above the river valleys. Their Mesozoic to Recent sediments (quartzite and limestone) and volcanics (rhyolite, basalt, and surficial lava) are in places intruded by plutons of granite and andesite. Mineralization occurs at contact zones and along fault lines, giving rise to silver, gold, and copper ores that have attracted miners since the seventeenth century (fig. 2).[1]

Channeled within the elongated valleys, five rivers and their tributaries drain La Serrana: the Río San Miguel on the west; followed eastward by the Río Sonora; the Moctezuma (or Oposura); and the Upper and Lower Bavispe in the northeast, both of which flow into the Río Yaqui, the larg-

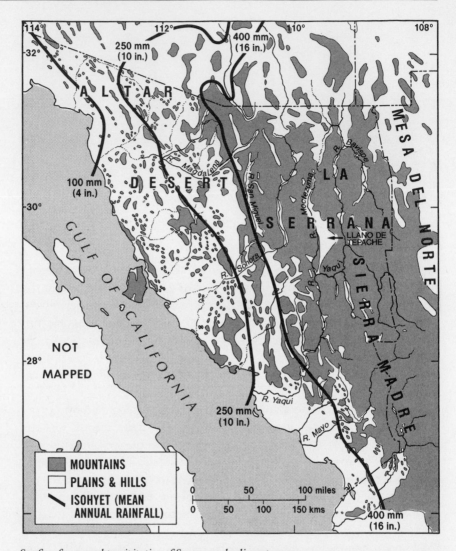

Surface forms and precipitation of Sonora and adjacent areas.

FIG. I

est of the Sonoran rivers. The narrow alluvial floodplains of most of these streams rarely exceed two kilometers in width; they usually form discontinuous stretches of arable land separated by narrow box canyons, or *cajones*, where the river has cut through consolidated volcanic ash or even granite (figs. 3 and 4). Each of the five main rivers is fed by springs and intermittent tributaries or arroyos that occur along their entire lengths.

The Sierra la Campanería, elevation 1,500–1,700 meters, between the Yaqui River and the Valley of Sahuaripa, is typical of the north-south ranges of La Serrana, eastern Sonora.

FIG. 2

Along the Yaqui and Mayo, however, only small pockets of alluvium afford opportunities for farming until the rivers break out of the mountains to form large delta plains along the coast of southern Sonora.

Flanking the floodplains along most of the rivers are remnants of thick gravel beds that rise twenty to one hundred meters above the Recent alluvium. Deposited by late Tertiary and Pleistocene rivers thousands of years ago, most of the gravel beds are sharply dissected; others are less so, however, and form flat-topped terraces or mesas overlooking the present valley (fig. 5). Free of flood, such terraces have served as habitation sites since prehistoric times, and today most of the river hamlets and pueblos, as well as the river roads, are found perched along them. As in the past, the north-south-trending valleys, some of them connected by transmontane roads and trails, form the main land routes of travel within La Serrana.

Other than the narrow river valleys, flattish surfaces within mountainous La Serrana are rare indeed. One extensive plain is the Llano de Te-

The Sonora River Valley near the village of Suaqui, eastern Sonora. Dissected Pleistocene terraces backed by north-south ranges border both sides of the river floodplain.

FIG. 3

pache (650–700 meters elevation) east of the Río Moctezuma. However, the widespread lava flows (*malpaís*) that cover most of its surface have precluded its use as an agricultural area of any importance.

The Río San Miguel marks the western boundary of La Serrana and roughly coincides with the mean annual isohyet (rainfall) of four hundred millimeters (16 inches) (fig. 1). Annual precipitation increases eastward with higher elevations, resulting in subhumid conditions near the western flank of the Sierra Madre Occidental. In general, the valleys receive less rain (four hundred to five hundred millimeters) than the higher mountain slopes and summits, where yearly precipitation may reach eight hundred to one thousand millimeters (thirty-two to forty inches).[2] Valleys thus support only arid to semiarid vegetation, in contrast to the adjacent oak-and-scrub-covered mountain slopes and pine and oak forests on crests over two thousand meters elevation.

Two periods of precipitation characterize the annual moisture regime of much of northwestern Mexico and adjacent portions of the American

The Sahuaripa River Valley (500 meters elevation) near Arívechi, eastern Sonora. A flat-surfaced terrace and north-south mountain range are seen in the background.

FIG. 4

Southwest.[3] Most rain falls in the summer months of July, August, and September, whereas less than a third as much comes in the winter period, from late November through early February. Severe drought occurs in late fall and late spring. Fed by moist air from the Gulf of Mexico, the "monsoon" rains of summer come as heavy, isolated afternoon thunderstorms, which sometimes cause flooding in the river valleys. Weak frontal storms from the Pacific bring in the light winter rains, locally called *equipatas*,[4] which often last for days over wide areas, with snowfall in the higher elevations of the Sierra Madre Occidental. Infrequently, tropical storms of subhurricane force that develop over the eastern Pacific Ocean in late summer and fall reach northwestern Mexico, causing disastrous floods that severely damage crops and erode valuable farmland in the valleys.[5] Along the coast of Sonora and Sinaloa such a storm is called "el cordonazo de San Francisco," or the lash of Saint Francis, since it occurs around the feast day (October 4) of that saint.[6] Despite occasional river valley floods in La Serrana, the duality of the rainfall season has permitted local farmers to

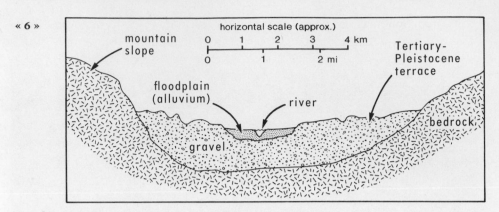

Stylized geomorphic cross section of a typical river valley in eastern Sonora.

FIG. 5

cultivate two crops annually, perhaps since prehistoric times. Although the major crops are grown during the summer rains, the scattered and variable nature of the thunderstorms often requires the use of artificial irrigation, as does the lesser rainfall of winter.

Because of its subtropical latitude (26°–32° N), the Sonoran Serrana enjoys cool to mild winters but hot summers. In the valleys daytime summer temperatures often reach above 40° C (104° F), whereas in winter mild (15°–20° C, or 60°–70° F) days prevail, but occasional light nighttime frosts are not uncommon. In the adjacent mountains temperature decreases with altitude, and in winter snow occasionally falls on the higher crests.

Western Sonora, or El Desierto

Geologically, Western Sonora constitutes part of the southwestern craton of North America, its dominant pre-Cambrian rocks having been formed an estimated 1.2 billion to 1.7 billion years ago; for much of that time the older granites and metamorphosed sediments, as well as the younger limestones and shales, have been exposed to atmospheric weathering.[7] Thus, the former mountain ranges of the area have been eroded down almost to their roots, forming isolated rock masses, or inselbergs, separated by wide, gently sloping *bajadas* composed of alluvial fans and pediments.[8] In places near the coast, Pleistocene and present-day rivers have created flattish deltaic plains. In the *bajadas* rich deposits of gold dust and nuggets, eroded from the former mountains, have formed placers, exploited since the late eighteenth century.

Western Sonora is also part of the most arid portion of North America. This dry zone, usually called the Sonoran Desert Region, which includes also the southeastern tip of California, southwestern Arizona, and much of Baja California, for most of the year lies under the influence of a high-pressure cell, the main cause of the excessive aridity. The northern part of western Sonora, called the Altar Desert, receives a mean annual rainfall of less than 250 millimeters (10 inches). In general, annual rainfall increases from the coast (100 millimeters, or 4 inches) inland toward La Serrana and southward from the Colorado River delta (50 millimeters or 2 inches) to Navojoa (380 millimeters or 14 inches), near the Sinaloa border.

As in La Serrana, in western Sonora precipitation is seasonal, with half to two-thirds of the annual total falling in summer convectional thunderstorms, the rest as light rains in winter. But the desert rainfall is far more variable than in La Serrana; some places along the coast may not receive rain for more than a year; other areas not too far distant may experience a single storm that brings more rain than the annual mean.[9]

Except the Colorado on the extreme northwestern margin of Sonora, the larger rivers that cross the desert toward the Gulf of California originate in La Serrana, but often their lower reaches are intermittent; their waters usually disappear in the desert gravels and flow to the Gulf only during summer floods. Until recently, such were the characteristics of the Magdalena-Asunción and Sonora rivers in the north. Since Pleistocene times both have built extensive delta plains of gravel and silt with fertile topsoil near the coast; they have also deposited copious amounts of groundwater, forming aquifers deep below the desert surface. But only in the last few decades have the soils and aquifers been exploited for agriculture (fig. 6).

In the southern portion of the desert, near the Sinaloa border, the Yaqui and Mayo rivers also have built extensive deltas near the coast. The Yaqui, which drains much of La Serrana, is the larger, but in the lower portions of both streams water flowed into the Gulf and formed wide natural levees that were cultivated by aborigines in pre-Spanish times. Today the water of both rivers is stored in large reservoirs and led off in canals for irrigation. No longer do the rivers reach the sea, but traces of former channels abound on the seaward side of the deltas and indicate the shifting courses of the distributaries that have formed the alluvial plain.

Practically all other desert streams in western Sonora are intermittent—dry, shallow beds for most of the year, carrying water only as flash floods after heavy summer downpours. Most of the water is not channeled, but moves off along the ground as sheet wash, eroding bedrock to form smooth, sloping pediments and depositing on them a thin veneer of

The delta plain of the Río Sonora (Costa de Hermosillo), most of which today has been converted to irrigated agriculture, as seen in the far distance. Photo taken from one of a group of inselbergs (Siete Cerros) 40 kilometers west of the city of Hermosillo.

FIG. 6

sand and gravel.[10] In the heat of summer most of the rainwater evaporates before flowing any distance, and little appears to penetrate the ground to help supply the water table.

Thus, except along the larger rivers, fresh water has always been at a premium in western Sonora. Natural water holes, or *charcos*, in clay flats or in intermittent stream-beds depend wholly on rainfall and are ephemeral at best; more dependable are the *tinajas*, or rock tanks—depressions in solid rock at the base of isolated mountains—which are fed by rainfall and in some places by springs. These hold a sizable volume of water, which may survive rapid evaporation for much of the year. In the desert the *tinajas* have been crucial water sources for humans and animals for centuries.[11]

Today most of the delta of the Colorado River lies within Baja California, and only the eastern third forms the northwesternmost part of Sonora. But for perhaps ten thousand years westerly winds have carried large quantities of fine deltaic sediment eastward, forming extensive areas of sand dunes that cover nearly eight thousand square kilometers in the northern part of the Altar Desert. Locally called *el gran desierto*, this "sand sea" is the largest of its kind in North America. In places the blowing sand has even engulfed the lower slopes of isolated mountains.[12] Largely bare of vegetation, the areas of active dunes present one of the most desolate parts of Sonora—a true *despoblado*, devoid of human habitation.

Especially during prehistoric and colonial times, the excessive heat of summer compounded the difficulty of human life in arid western Sonora. Immediately prior to the break of the scattered monsoon rains in late June or early July, daytime temperatures often rise to 45° C (120° F). After gold was discovered in the Altar Desert late in the eighteenth century, mining activity ceased during those months in part because of the enervating heat. Winter temperatures are mild, nighttime frosts are infrequent, but on rare occasions in January or February cold waves push in from the Great Basin of the United States, causing temperatures to drop slightly below freezing for short periods.

Sonoran Vegetation

Despite prevailing aridity, the natural vegetation of western Sonora, like that of the rest of the Sonoran Desert Region, is surprisingly lush and varied, to the extent that the term "arboreal desert" is commonly applied to much of the plant cover. According to botanist Forrest Shreve,[13] "the Sonoran Desert Region is by far the richest in number of life forms and in variety and development of communities" of all the North American deserts. A myriad of species, including grasses, shrubs, low trees, and many kinds of cacti, such as the tall columnar varieties, as well as several types of century plants, or agaves, characterize the vegetation of most of the area (fig. 7). This anomaly may be due in part to low elevation and the dual precipitation regime described earlier. The light winter rains, especially, coming during the period of least evaporation and penetrating the soil more deeply than do the flash downpours of summer, likely maintain plant life more effectively than otherwise would be the case.

Arboreal desert vegetation near the coast southeast of Guaymas. Cactus, including the columnar organpipe and the branching cholla (foreground) and low trees such as ironwood and palo verde, typify the plant cover of western Sonora, especially in the delta plains, which receive from 250 to 300 millimeters of annual rainfall.

FIG. 7

Plant variety, density, and size increase with greater precipitation. In the driest part of the Altar Desert, the north (one hundred to two hundred millimeters yearly rainfall), scattered low shrubs such as creosote bush (*Larrea tridentada*) and burro bush (*Franseria dumosa*) prevail. Farther inland and southward, in areas of greater precipitation (two hundred to four hundred millimeters), one finds high shrubs and low trees, mainly legumes, such as various acacias (*Acacia*, spp.), mesquite (*Prosopis*, spp.), the yellow-flowering palo verde (*Cercidium*, spp.), and the durable ironwood (*Olneya tesota*). Other shrubs and low trees include the spindly red-flowering ocotillo (*Fouquieria*, spp.), and the copal tree (*Bursera*, spp.), among many others. Mesquite together with stunted cottonwood (*Populus fremontii*) and occasional willows (*Salix*, spp.) are found as dominants along intermittent watercourses where moisture is available.[14]

The most spectacular forms are the cacti, especially the columnar varieties that one usually associates with the American Southwest. The familiar tall sahuaro (*Cereus* [or *Carnegiea*] *gigantea*) extends from Arizona to near the Sonora-Sinaloa border, as does the smaller organpipe, or pitahaya (*Stenocereus* [or *Lemaireocereus*] *thurberi*). The largest of all columnar types is the cardon (*Pachycereus pringlei*), fifteen to twenty meters tall, found mainly in Baja California but also near the Gulf coast of Sonora. Smaller cacti include various kinds of prickly pear and cholla (*Opuntia*, spp.) and the hedgehog cactus (*Echinocereus*, spp.).

Still another important desert plant is the agave (*Agave*, spp.), often called the century plant because of its long life before flowering. These low plants have thick, fleshy leaves and a flower stock that develops after ten to twelve years. In Sonora and elsewhere in northern Mexico they grow on dry, rocky hillsides and are used for food and fiber.

Curiously, along rocky intermittent stream courses in the mountains of western Sonora one occasionally encounters palms, among them a species of the tall *Washingtonia* of southeastern California (*W. sonorae*), which occurs in places along the Gulf coast as far south as Guaymas. Smaller fan palms, such as *Erythea* and *Sabal*, spp., are found in similar environments within the mountain canyons of La Serrana.[15]

In southernmost Sonora and extending into Sinaloa a vegetation association, called thorn forest, occupies the lowland plains, including the Yaqui and Mayo river deltas. Many species of acacia, mesquite, and cactus (especially the columnar pitahaya), as well as low tropical trees, such as torote (*Bursera adorata*), brazil (*Haematoxylon brasiletto*), tree morning glory (*Ipomea aborescens*), and pochote, or kapok (*Ceiba acuminata*), grow to heights of 6.5 meters or more. Farther inland, within the foothills of up to 900 meters elevation, such plants increase in height with greater rainfall, and in the canyon bottoms large trees, including the native fig (*Ficus*, spp.), bald cypress, or sabino (*Taxodium mucronatum*), and tall *Erythea* palms dominate the streamside plants, all forming a subtropical to tropical deciduous-evergreen association.[16]

Many of these desert plants have been important for food, building materials, and tools to the people of Sonora. Of particular significance have been the leguminous plants, such as mesquite, the pods of which, rich in protein and carbohydrates, are ground to a meal for making bread and porridge. The fruits of various cacti, mainly the pitahaya (which is an excellent source of ascorbic acid, is palatable, and is easily gathered), are still utilized today, especially by the Seri Indians of western Sonora and the Yaqui and Mayo in the south. Again, the leaves and hearts of the agaves were gathered and cooked in shallow pits lined with hot stones to prepare a nutritious food called mescal. In the Serrana of eastern Sonora

Oak and mesquite grassland, northeastern Sonora, near Cananea, 1,400 meters elevation.

FIG. 8

today agave hearts are gathered to make an illicit distilled liquor, the mescal of Bacanora.[17] Moreover, ironwood has been the source of material for fashioning utensils for home use, and the leaves of palms are still gathered for thatching roofs, and in some areas the trunks are cut for constructing house walls.

Many plants characteristic of the western desert occupy an important place in the ecological zones within the valleys of La Serrana in eastern Sonora. In fact, on his map of the Sonoran Desert Region, Shreve extends the desert plant assemblage into many of the river valleys of La Serrana.[18] Mesquite, cottonwood, and willow form part of the woodland along rivers and lower sections of tributary arroyos. On the adjacent terraces or mesas, small trees such as ironwood and palo verde, thorny shrubs mainly of the genus *Acacia*, cacti such as organpipe and cholla, and various agaves and grasses form a desertlike assemblage. In north-central and northeastern Sonora near the United States border, where annual precipitation exceeds four hundred millimeters, grasses of many species abound and, mixed with scattered stands of mesquite and oak, present a steppelike landscape (fig. 8). These grasslands have supported herds of livestock since the beginning of Spanish settlement in the mid-seventeenth century. Among the grasses one of the most valuable for grazing is the black grama (*Bouteloua*, spp.), eighteen species of which have been collected in the Bavispe area of northeastern Sonora.[19] Green and lush after the summer rains, this grass dries to a palatable, nutritious hay during the dry season, thus affording year-round forage.

On the higher mountain slopes of La Serrana, where rainfall is more abundant, an evergreen woodland composed mainly of oak (*Quercus*, spp.), juniper (*Juniperus*, spp.), and pinyon (*Pinus edulis*) gives a distinct appearance to the Serranean landscape, as do the pine and oak forests on the high mountain crests (fig. 9).[20]

Sonoran Wildlife

Both the Sonoran Desert and the Serrana must have been well populated by wildlife before and after the Spanish Conquest. Being mobile creatures and adaptable in their ecologic tolerance, most animals may range widely across various climatic and vegetational zones.[21] Thus, in Sonora, except for some species that became restricted to particular natural conditions, in general, the same faunal assemblages are found in both the low, arid western and the more humid and mountainous eastern portions of the state. For the aborigines wild animals were important sources of food, of skins for various uses, and of items employed in ceremony. The coming of the Spaniards with their livestock led to the gradual attrition of wild predators and competition between grazing and browsing species. Modern hunting methods with rifle, steel trap, and poison further reduced the numbers of native animals, so that today Sonoran wildlife must represent only a fraction of its pre-Spanish population.

In terms of human use, the larger mammals, particularly the ungulates, represent the more important wildlife in Sonora, as elsewhere in North America. Two species—the mule deer (*Odocoileus hemionus*) and the common white-tailed deer (*O. virginianus*)—provided the favorite meat consumed by Sonoran Indians, and the hides were prized for making footwear, sleeping mats, and other household objects; among the Yaqui and Mayo Indians the dried head and antlers served as important paraphernalia in religious ceremonies. Although both species browse throughout

Open oak forest, east-central Sonora, on the western flank of the Sierra Madre Occidental, 1,800 meters elevation.

FIG. 9

northwestern Mexico, the mule deer is more at home in the Sonora Desert Region, the white-tailed deer, in the better-watered valleys and mountains of La Serrana. Other ungulates hunted less frequently by aborigines for food included the pronghorn antelope (*Antelocapra americana*), bighorn sheep (*Ovis canadensis*), and the collared peccary, or javelina (*Tayassu angulatus*). Smaller mammals, such as the jack-rabbit (*Lepus*, spp.) and the cottontail (*Sylvilagus*, spp.), were hunted and trapped, but these and numerous rodents, such as gophers, field mice, squirrels, raccoons, and coatis, are considered today more as crop-damaging pests than as sources of food.[22]

Carnivorous predators—the cats, such as the common bobcat (*Lynx rufus*), the North American cougar, or mountain lion (*Felis concolor*), and the neotropical jaguar (*F. onca*) and ocelot (*F. paradalis*), and the black bear (*Euarctos americanus*)—are encountered mainly in the mountains of

La Serrana. The ubiquitous coyote (*Canis latrans*) lives on small game throughout western North America, including Sonora, but the gray wolf (*C. lupus*) and the gray fox (*Urocyon cinereoargenteus*) confine their hunting to La Serrana, while the kit fox (*Vulpes macrotis*) is mainly a desert animal.[23]

Freshwater fish, such as some species of catfish and cichlids, live in the streams of La Serrana, but they have not been an important food source for local inhabitants. During dry periods piscicides derived from certain plants have been used to stupefy fish in shallow riverbed pools. Also along streams various kinds of wild ducks and other water birds were probably hunted by aborigines, but more important for food were quail, doves, and the wild turkey (*Meleagris gallopavo*) that inhabit the bush-covered hillsides.[24]

Along the Gulf of California coast, shore birds abound: seagulls, cormorants, herons, snowy egrets, terns, ospreys and many others, but the most important is the pelican (*Pelecanus occidentalis californianus*), which the Seri Indians caught for food. Offshore sea mammals, such as the sea lion (*Zalophus californianus*); a reptile, the green sea turtle (*Chelonia mydas carrinegra*); and many kinds of fish are still exploited by the few remaining Seri of Sonora.[25]

CHAPTER 2

The Aboriginal Cultures of Sonora

As in other parts of Mexico, in Sonora the nature of aboriginal culture influenced to a large degree the directions that Spanish conquest and settlement would take. Obviously, the status of the aborigines of any area settled by an invading people may set the stage for subsequent development of the geographical scene as well as for the evolution of economic and social policies on the part of the invader. The greater the population and the higher the culture of the natives, the stronger their influence may be. This phenomenon is especially true for Hispanic America, where in most of the areas settled by Spaniards the Indian element was strong during the colonial period and is still important, despite centuries of acculturation.

In pre-Conquest times Sonora was inhabited by several tribes speaking different but often related languages and practicing economies that ranged from primitive hunting and gathering to relatively sophisticated farming with irrigation (fig. 10). Most of the languages spoken belonged to the Uto-Aztecan stock, widespread in Mexico. In southern Sonora these included Cáhita, represented mainly by the Yaqui and Mayo dialects; in the eastern and central parts of the state, Opata (including the Eudeve dialect) and Pima Bajo; and in the northwest, Pima Alto and closely related Pápago. However, the Seri, along the Gulf of California coast, spoke a Yuman tongue, related to that of the desert tribes inhabiting Baja California and western Arizona; and the Apache tribes on the northeastern border of Sonora spoke Athapascan languages that were more common in western Canada and in Alaska.

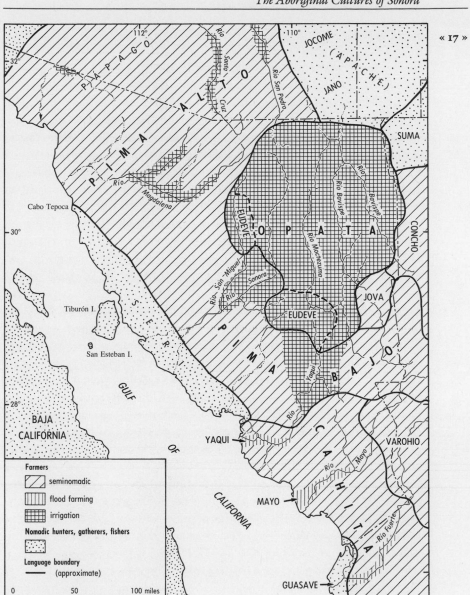

Aboriginal languages and economies, ca. A.D. 1500. Modified from Sauer, 1935a (map: Tribes and Subsistence Modes of the Aboriginal Population of Northwestern Mexico).

FIG. 10

The Farming Cultures

Most of eastern Sonora was occupied by agricultural Indians, representing the expansion of a farming culture into northwestern Mexico, possibly from the central part of the country. Among the farmers were the Opata and some Pima Bajo, who practiced canal irrigation along the larger rivers of La Serrana. The Opata appear to have been the more numerous and culturally superior of the Sonoran aborigines, as evidenced by reports of sixteenth-century Spanish explorers who cited the Opatas' abundant crops, large pueblos, and multistoried houses. During their epochal trip through what is now Texas, New Mexico, and northwestern Mexico, Cabeza de Vaca and his companions in 1537 encountered people cultivating abundant crops of maize, beans, and squash in the valleys of northeastern Sonora. Three years later the Coronado expedition on its trip northward into the Great Plains passed through Opata country, where sizable towns and abundant harvests were noted.

Such reports, formerly dismissed by most scholars as exaggerations, have now been partially substantiated by recent archaeological investigations. The most detailed survey was carried out along the middle portion of the Sonora River Valley, probably the main center of the Opata people. Directed by personnel from the Department of Anthropology, University of Oklahoma, and the Hermosillo branch of the Instituto Nacional de Antropología e Historia in 1977 and 1978, this survey revealed a relatively advanced aboriginal culture.[1] On the gravel terraces overlooking the cultivated plain, some two hundred sites were excavated along a fifty-one-kilometer stretch of the valley. These indicated the presence of house foundations ranging in number from nine to twenty-five for hamlets, twenty-six to one hundred for small nucleated villages, and over one hundred for larger towns. The two largest settlements, which the archaeologists termed "regional centers," had over two hundred house or building foundations. From this hierarchy of settlement sites it was suggested that two "statelets" had developed in the surveyed portion of the Río Sonora Valley, each dominated by a large regional center that served for centralized functions. From an analysis of pottery and remains of other artifacts found in the sites, it was thought the statelet political and social pattern may have evolved between A.D. 1350 and 1550.[2]

Constructed of river-worn stones, most of the house foundations were rectangular, and the presence of low heaps of adobe melt nearby suggested that house walls were of sun-dried brick and possibly covered with flat roofs of woven mats and dirt, as described in early Spanish accounts. Only one definite archaeological indication of the practice of irrigation in the valley was found: a petrograph interpreted as a crude map possibly showing streams, canals, and settlements.[3] Other evidence of pre-Spanish irrigation comes from early accounts of explorers and priests and from the assumption that present-day methods of irrigation may have been developed in pre-Conquest times. Along the Sonora River Valley an earlier phase of occupation, radio-carbon dated between A.D. 100 and 1200, was evidenced by the remains of semisubterranean structures (termed "pit houses") and the presence in adjacent lower mountain slopes of terraced conical hills (called *trincheras*).

Another detailed archaeological survey was conducted along the upper portion of the San Miguel River Valley, immediately west of the Río Sonora.[4] At Spanish contact this area was inhabited by an enclave of Eudeve-speaking people, whose language was closely related to Opata. The San Miguel presented an ancient settlement pattern similar to that of the Sonora Valley, with rectangular house foundations on Pleistocene gravel terraces above the flood-plain. Likewise, an earlier phase was characterized by *trincheras*, or fortified hills near mountain slopes; such sites extend into the lower part of the valley, where rectangular house foundations are absent.

Only reconnaissance archaeological surveys have been made along other river valleys of La Serrana that were occupied by either Opata or Pima Bajo. In the early 1930s Carl Sauer on one of his several field excursions into Sonora discovered rectangular house foundations on terraces overlooking the town of Arívechi along the Sahuaripa River, a tributary of the Yaqui. Similar finds were encountered along the Moctezuma (Oposura) River near Batuc, where foundations measured eight by twelve feet.[5] In 1927, Monroe Amsden had found such house bases on the Upper Bavispe River, near Colonia Oaxaca, as well as along the Moctezuma and Sonora rivers.[6] From such evidence he coined the term "Río Sonora Culture," now applied to the late prehistory of much of the eastern part of the state. These and other more recent reconnaissance surveys have led some archaeologists to surmise that the entire Opatería consisted of statelets that developed within most of the north-south-trending valleys of La Serrana, similar to the units inferred for the Río Sonora Valley.[7] However, before such a political and social pattern can be firmly established, detailed surveys must first be made in all of these river valleys.

Judging from the archaeological surveys and the statements made by early Spanish explorers, the Opata country of northeastern Sonora appears to have been the most densely populated area of northwestern Mexico at

« 20 »   Spanish contact.[8] Doolittle conservatively estimates the population of just the upper Río Sonora at ten thousand, but puts the population of the entire Opatería at one hundred thousand,[9] much more than Carl Sauer's earlier estimate of sixty thousand, which at the time of publication (1935) was considered excessive by many anthropologists. Riley and Pailes both suggest early trade contacts between the Opatas and the people of Casas Grandes, an emporium having commercial ties with the Valley of Mexico.[10] Exchange of cotton cloth from the Opatería for Casas Grandes ceramics, sherds of which are found at many sites in northeastern Sonora, is suggested as a possible basis for trade. Doolittle, however, would favor local intervalley trade over Casas Grandes influence as an impetus for the advanced social and political status of the Opata people.[11]

By the seventeenth century, when the Opata-Eudeve area of La Serrana was missionized by the Jesuit order and invaded by Spanish miners, the relatively advanced native culture suggested by archaeology and reported by early Spanish explorers had declined. Through causes unknown (disease?), population may have decreased, the statelet system of sociopolitical organization was simplified, and town settlement seems to have been reduced to villages and hamlets scattered along the river valleys. Agricultural techniques, however, including canal irrigation, apparently were retained intact.[12]

The numbers of Cáhita-speaking Indians, most of whom lived along the coastal plain and river valleys of Sinaloa, also were large. In southernmost Sonora the Cáhita were represented by the Yaqui and Mayo tribes, whose agriculture, unlike that of the Opata, was based not on canal irrigation, but on natural flood farming. In southern Sonora this system relied on the annual or semiannual flooding of the Yaqui and Mayo rivers within their large delta plains near the Gulf of California coast and on the planting of crops in the moist soil of the natural levees as the floods receded. Fed by tributaries that head in the Sierra Madre Occidental, both rivers flood twice annually, the maximum flow occurring during the heavy summer rains, a lesser flooding in February and March, after the light winter rains and snowmelt in the sierra. Hence, the Yaqui and Mayo were able to harvest two crops yearly, the major one of maize, beans, squash, and cotton gathered at the end of the summer rains.[13] Pérez de Ribas, one of the first Jesuit missionaries to enter southern Sonora early in the seventeenth century, estimated the population of the Yaqui and Mayo at thirty thousand each, the former living in some eighty *rancherías*, or hamlets, on both banks of the river within the delta plain.[14] Both men and women planted crops with a dibble shaped like a spade. Women used horizontal ground «21» looms to weave cotton cloth. During periods between maize harvests and in times of crop failure both sexes gathered fruits of wild plants: mesquite pods, pitahaya and other cactus fruit, and agave leaves and hearts, which were cooked in pits to make the nutritious mescal. Caught in traps placed at the mouths of estuaries along the Gulf coast, sea fish formed an important part of the Yaqui diet, as did wild animals, such as the white-tailed deer and the peccary, hunted in the scrub-covered hills north and east of the delta plain.[15] Spicer has suggested that the Yaqui obtained possibly half of their food from gathering, hunting, and fishing.[16] Less information is available for the Mayo at Spanish contact, but their economy was likely similar to that of the Yaqui.

The Pima-speaking people, who bordered the Opata country on the south, west, and northwest, were probably much fewer in numbers than their more sophisticated neighbors. The terms "Bajo" (Lower) and "Alto" (Upper) are geographical, not linguistic, for the people in both categories spoke the same language with only dialectical differences.[17] Apparently, the terms were introduced by Jesuit priests during their initial evangelization activities in the seventeenth and early eighteenth centuries. Some investigators have suggested that an early invasion of Opata speakers into formerly Pima territory may have been the cause of the geographical Opata-Pima language pattern at Spanish contact.[18]

Pima Bajo extended from the middle Río Sonora around Ures eastward to the middle portion of the Río Yaqui and thence into the Sierra Madre Occidental as far as the present pueblos of Moris and La Junta in western Chihuahua (fig. 10). Westward from the plain of Ures they lived along the lower part of the San Miguel River and may well have settled along the Río Sonora as far downstream as present-day Hermosillo. They also may have hunted in the Sonoran Desert, occasionally making contact with Seri nomads to the west and the Pima Alto farther north. Pennington indicates that the heartland of the Pima Bajo lay along the central portion of the Río Yaqui, with *rancherías* on both sides of the river, from which were later formed the mission towns of Onavas, Movas, and Cumuripa.[19]

Another important center was the Ures plain along the Río Sonora. In both areas rain-fed agriculture probably predominated, but "hand irrigation," which involved the carrying of water in clay pots from the river to nearby fields of maize, beans, and squash, may have been practiced, as it is today. Basing his opinion on lack of favorable terrain, Pennington discounts the early Spanish reports of canal irrigation by the Pima Bajo along

«22» the Río Yaqui;[20] however, canals may have been constructed in the Ures plain, thought to be the site of "Corazones," which the early Spanish explorers praised as a veritable garden of maize and other native crops.

Unfortunately, few archaeological surveys have been undertaken in the Pima Bajo area of Sonora, so that population estimates of those people at Spanish contact are difficult at best. Sauer's estimate of twenty-five thousand has been seriously questioned by Pennington, whose guess of only six thousand is based on his meticulous fieldwork in the area and close study of Jesuit reports on mission head counts of Indian converts.[21] His figure, however, may be too low, as he himself admits.

Living in the more arid sectors of northwestern Mexico and southern Arizona, the Pima Alto were often thought to be culturally inferior to their Pima Bajo kin.[22] But as part-time farmers, they, like all of their agricultural neighbors to the south, were at the mercy of available water from rain or stream. At Spanish contact their total population may not have exceeded fifteen thousand.[23] Linguistically, they comprised several dialect groups of Piman. The Pápago inhabited the most arid part of northwestern Sonora and southwestern Arizona; farther east, the Soba farmed land along the lower Magdalena and Altar rivers; the Himiris, along the upper Magdalena; and the Sobaipuri, the largest group, lived along the San Pedro, Santa Cruz, and middle Gila rivers, mainly in what is now southern Arizona. After Spanish conquest all of these areas came to be known as the Pimería Alta, the northernmost frontier of colonial Sonora.

The Pápago, or Desert Pima (the current preferred name is Tohono O'odam), had meager possibilities for farming in their severe arid environment. The "Sand People," those living near the dunes east of the Colorado delta, may have been nomadic hunters and gatherers without agriculture.[24] Most of the Pápago, however, were semi-nomadic, planting and tending their small, scattered fields during the summer rains; the rest of the year they wandered about, the men hunting, the women gathering edible plants. These part-time farmers utilized natural flood irrigation, but they could not depend on the annual or semiannual floods that the Yaqui and Mayo enjoyed along their rivers; instead, the Pápago were restricted to the unreliable flash floods that occurred in small arroyos after sporadic thunderstorms. Rushing down canyons through low mountain slopes, the floodwaters spread out in broad sheets over gently inclined alluvial fans. The Indians quickly planted their crops in the moistened thin soil. In places dikes and ditches were constructed to control and direct water as it issued from the canyon mouths. In some years the rains failed, and the natives were reduced to hunting and gathering; occasionally, they must have suffered famine.[25] In only one place is there evidence that the Pápago used canal irrigation: at Sonoíta, near the present Sonora-Arizona border, where permanent water was available in an arroyo (Arroyo Sonoíta) from which ditches were made to lead water onto the fields.[26]

Canal irrigation was common among the Riverine Pima—the Sobaipuri—who lived in permanent villages along the narrow alluvial floodplains of the San Pedro, Santa Cruz, and a short stretch of the Gila rivers in southern Arizona. During the summer rains these rivers were full-flowing, and ditches were used to divert water to the fields, as described by Padre Kino and Captain Mange late in the seventeenth century, when these Indians were being evangelized by the Jesuit order. At that time the Spaniards reported a population of two thousand along the San Pedro, twenty-five hundred along the Santa Cruz. On the latter stream the village of Bac (mission San Javier Bac, near present Tucson) was the larger settlement, and springs supplemented the summer rains for canal irrigation.[27]

Less is known about the farming practices of the Himiris and the Soba Pima, who lived along the Magdalena and Altar rivers in northern Sonora, but since these streams were full during the summer rains and usually intermittent during the winter months, it has been assumed that canal irrigation was practiced along them for summer crops. On a trip down the Magdalena in 1694, Captain Mange reported irrigation by the Indians at Caborca, on the middle section of the river.[28] As many as four thousand Soba Pima have been estimated to have lived along the middle Magdalena and Altar rivers at Spanish contact.[29]

The provenience of irrigation techniques among the Pima Alto is unknown. It is doubtful that they stemmed from the ancient and sophisticated irrigation practiced by the Hohokam culture along the Gila and Salt rivers of Arizona. The Hohokam people were constructing an extensive canal system as early as A.D. 800, development reaching its maximum by A.D. 1200–1400. Thereafter, the canals were abandoned, and the crude ditches used by the Sobaipuri Pima along the Gila at Spanish contact were a far cry from the large, well-made canals of the Hohokam period.

Like all other aboriginal farmers of Sonora, the Pima Alto cultivated the usual Mesoamerican crop triad of maize, beans, and squash, plus cotton and perhaps grain amaranth. Although grown by most of the Sonoran farmers, the tepary bean (*Phaseolus acutifolius*) held a special place among the Pima Alto. They also cultivated the common Mexican pinto bean (*Ph. vulgaris*), but the tepary, high in protein and adapted to an arid environment, is one of the few domesticated beans that extends into truly desert areas.[30] According to Nabhan, the Pima Alto still consume more beans

« 24 » than any other Indian group in Mexico, and historically their diet has been based on legumes rather than on grains or meat.[31]

The Pápago especially, who concentrated on the cultivation of the tepary bean, also gathered pods and seeds of wild legumes to supplement their diet. These plants included the wild tepary, the mesquite, the pods of which were consumed in the form of ground meal, and the seeds from the pods of the palo verde (*Cercidium*, spp.) and ironwood (*Olneya tesota*), which were parched and eaten dry or mixed with water.

Aside from work on Hohokam culture in Arizona, archaeology in Pimería Alta to a large degree has focused on the curious sites called *trincheras*, terraced hillside habitations constructed on steep mountain slopes and often on conical hills that overlook arable land along streams. Such sites are most abundant in the Magdalena-Cocóspera river basin and the Altar Valley in north-central Sonora, but others occur along the San Miguel, Sonora, and Santa Cruz river valleys and as far north as the Sells Pápago Reservation in Arizona. Although these sites have attracted the attention of travelers since the eighteenth century, Sauer and Brand made the first serious archaeological reconnaissance survey of them in 1928 and 1930.[32] Most of them have been regarded as defense sites, possibly places of refuge for local inhabitants when their villages and fields were menaced by invading enemies. Spaces between the rock-walled terraces usually contain potsherds, remains of grinding stones, and even small house foundations and cooking pits, perhaps indicating lengthy periods of siege.[33]

The defense theory, however, is uncertain. Dated between A.D. 800 and 1100, most of the *trincheras* sites belong to a relatively early phase of settlement that corresponds to the beginning of Hohokam culture farther north. If the defense theory can be substantiated, the "Trincheras Culture" may represent a long period when agricultural areas in northern Sonora experienced warfare between local tribes or were periodically raided by nomadic desert people from the southwest or the northeast, just as the same area was attacked and partially overrun by nomads much later, during the Spanish colonial period.

The Hunting and Gathering Cultures

At the time of Spanish contact, two widely separated hunting and gathering groups represented the lowest aboriginal culture within the Sonoran area: the Seri along the central Gulf of California coast, and the several Apache-related tribes that bordered the Opatería in present-day Chihuahua and New Mexico. Although their subsistence level was rudimentary, both would become a significant negative factor for Spanish settlement in much of Sonora during the colonial period.

With a contact population of probably no more than five thousand, the Seri bands extended along the central desert coast from Guaymas northward to Cabo Tepoca, including the offshore islands of Tiburón and San Esteban. Since they spoke the only Yuman tongue in Sonora, the Seri may well have migrated across the Gulf from Baja California, where several languages belonging to the Yuman family were spoken. The presence of many shell mounds along the Sonoran coast may indicate accumulation by early Seri migrants.

The Seri who camped along the immediate coast and on the offshore islands were primarily fishers who caught finfish (mainly croaker and sea trout) and green turtles from their balsa reed boats and trapped sea birds, such as pelicans and gulls.[34] Farther inland other bands gathered fruits from a large assemblage of desert plants. Mesquite pods, ground on stone metates and consumed as a gruel, were the main source of protein, while fruits of the pitahaya, sahuaro, and other columnar cacti furnished essential vitamins and served as an antiscorbutic. Starch for the diet came mainly from the pit-cooked fleshy leaves and hearts of agave plants, which grew in abundance in the Sonoran Desert.[35]

Although ignorant of any kind of agriculture, the Seri were good potters and excellent basket makers; they were also adept at fashioning various kinds of tools from the hard ironwood tree. The few remaining Seri (some 250 individuals) who live on the coast near Desemboque still reject farming but gather, fish, and hunt, as did their ancestors. They have accepted only a minimum of Hispanic culture during their two hundred–year period of contact. Women still weave beautiful reed baskets and men carve handsome objects of ironwood, both products eagerly sought by tourists and museum personnel. Perhaps because they represent one of the few functioning aboriginal band-level hunting and gathering societies in North America, the remaining Seri have been studied and reported on ad infinitum by ethnologists and historians.[36]

The Apache hunters and gatherers, who became a serious menace to the Spanish colonists and native sedentary Indians in northern Sonora during the colonial period, were late-comers, arriving in southwestern New Mexico and northwestern Chihuahua apparently during the last decades of the seventeenth century. Athapaskan-speaking, Apache tribes for centuries had gradually been migrating southward from Canada into the

« 26 »   semiarid High Plains, where they became bison hunters.[37] In 1541 members of the Coronado expedition encountered Apache (probably of the Lipan branch), whom the Spaniards called "Querechos," in the Texas Panhandle. Subsisting mainly on bison, antelope, and deer and using dogs for transport, these Indians occasionally raided Pueblo villages in the upper Rio Grande Valley of New Mexico, but also traded hides for Pueblo pottery.[38]

By the 1680s an Apache group, known later as the Gileños, had reached the high grassy basins of northwestern Chihuahua, where they allied with the Suma, Jocome, and Jano tribes. The language of the last three is considered by some investigators as belonging to the Athapaskan family,[39] by others, as part of the Uto-Aztecan stock.[40] The combined forces began raiding Spanish stock ranches in the Chihuahuan basins, but eventually the marauders were expelled, resettling in the upper Gila River drainage. This area became the main base for the subsequent forays into northern and central Sonora that continued intermittently for two hundred years (see chapter 6).

CHAPTER 3

Spanish Settlement of Sonora: The Missions

Spanish settlement of Sonora did not occur for nearly a century after Francisco Vázquez de Coronado and his band made one of the first *entradas* through the area searching for the fabled Seven Cities of Cíbola. Early in the seventeenth century, the Jesuit order, which was bent on saving the souls of the natives for Christianity, began to establish missions in Sonora. By the 1640s Spanish lay settlers, mainly miners from the central plateau, had crossed the Sierra Madre Occidental into Sonora.

Initially, both forms of settlement were restricted mainly to La Serrana, or eastern Sonora, where the missions were confined to the river valleys and served sedentary and seminomadic Indian farmers, and the mines were established in the adjacent mountains for exploiting silver deposits. This juxtaposition of mine and mission gave rise to a long-lasting economic and social relationship between the priests and the lay Spaniards—a relationship that was usually complementary but could be contentious.

Jesuit Mission Activity in Sonora

Except for a brief flurry of evangelization by the Franciscan order in northeastern Sonora during the mid-seventeenth century, the Jesuits dominated mission activity in the province for more than 150 years, from the time of their entry among the Mayo and Yaqui (1614–1617) until the time of the expulsion of the order from the Spanish colonies in 1767 (fig. 11). The Jesuits approached Sonora from the province of Sinaloa, where the northern frontier of permanent Spanish settlement along the Pacific lowlands had languished near Culiacán since the mid-sixteenth century. In 1590 a few Jesuit priests began work among Cáhita-speaking Indians along the

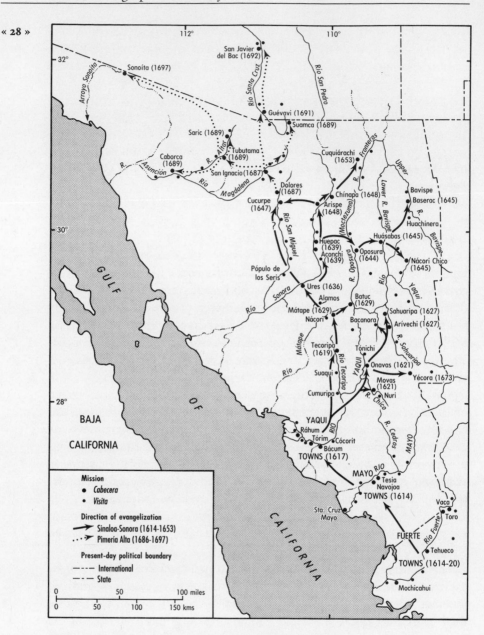

Jesuit evangelization of Sonora, seventeenth century.

FIG. II

Sinaloa River north of Culiacán, and the following quarter century saw a gradual extension of missions into the densely populated Cáhita territory of northernmost Sinaloa, including the deltas of the Fuerte, Mayo, and Yaqui rivers.[1]

In 1614 the Jesuits were successful in consolidating the numerous native *rancherías* (small, scattered hamlets) within the Mayo delta into seven larger mission settlements along the river.[2] These included the more important pueblos of Tesia, Navojoa, and Santa Cruz de Mayo, which were made *cabeceras*, or church centers, of two thousand to three thousand population, each with a resident priest; the four smaller pueblos were made *visitas*, each under the jurisdiction of a *cabecera* and attended normally by an itinerant missionary. For the Mayo Indians the newly founded *cabeceras* became sacred ceremonial centers, which the present-day Mayo descendants still venerate, if only in legend.[3] The *cabecera-visita* system of mission settlement was employed by the Jesuit order throughout Sonora and by other orders elsewhere in Spanish America. A group of consolidated settlements that included a *cabecera* and its various *visitas* was usually termed a *partido*.[4] The consolidation, or "reduction," of scattered native habitations was thought necessary to ensure proper religious and social control over the neophytes. A similar process had been used earlier by secular Spanish authorities in central Mexico, more to oversee tribute collections and native labor supply than to spread the Christian faith.

In 1617 the noted Jesuit priest Pérez de Ribas entered the Yaqui delta to convert the large Indian population living along the lower portion of the river. Despite a sharp encounter with the Spanish military shortly before, leaders of the bellicose Yaqui had actually asked for Jesuit missionaries to visit their river and to introduce the new religion to their people. It was Jesuit policy to create among Indian groups a strong desire to issue such "invitations," for the padres needed the cooperation of native leaders to aid in the conversion process and to carry out subsequent social changes. The Yaqui, having observed favorable aspects of missionary work among the neighboring Mayo, such as selection of politically prominent natives for places of authority within the local church and the introduction of new crops and domestic animals, were thereby "enticed" to invite priests into their area. The Jesuits employed this technique of "social seduction" in their missionary endeavors throughout Sonora.[5]

Pérez de Ribas had no difficulty understanding and speaking to the Yaqui people, for he had spent time in northern Sinaloa learning other Cáhita tongues, most of which were mutually intelligible. He thus had abided by another Jesuit rule that required missionaries to learn and to preach in the native language of the area to which they were assigned. Usually rigorously enforced, this rule may have been one of the most important reasons for the success of the Jesuit missions in Sonora.[6]

At the time of his entry into the Yaqui area, Pérez de Ribas estimated that some thirty thousand Indians were living in about eighty *rancherías* scattered within the river delta. By 1623 he and his successors had persuaded the native leaders to congregate a large portion of the population into eight pueblos of twenty-five hundred to three thousand inhabitants, each with a substantial church of stone and adobe. Moreover, in addition to the Jesuit religious structure, each pueblo had its secular political organization, with a native governor and another leader as mayor or judge (*alcalde*), both elected by the townspeople. The pueblos nearest the gulf, Belém (or Beene) and Huirivis, were located along the then-active northernmost river distributary, which was abandoned probably in the eighteenth century; its former channel is hardly discernible today. The other towns were located on the left, or south, bank of the river—Ráhum, Pótam, Vícam, Tórim, Bácum, and Cócorit. In the eyes of the native inhabitants these eight pueblos became sacred places, and still play a special symbolic role in Yaqui culture. Among other factors, they were instrumental in giving the Yaqui people a strong sense of tribal unity and place.[7] But today, Bácum and Cócorit are occupied mainly by Mexicans, and at least one Yaqui pueblo has been moved to the right, or north, bank of the river, their lands on the south bank having been acquired by Mexicans and other foreigners. Present-day Belém and Huirivis are now practically deserted and no longer occupy their original sites.[8]

By 1685, and perhaps earlier, the *cabecera-visita* structure was applied to the Yaqui missions: Ráhum, Tórim, and Bácum were designated as *cabeceras*, the others as *visitas*. For administrative purposes all of the missions within the Yaqui and Mayo deltas were placed into a geographical unit termed a "rectorate"—the Rectorate of Nuestro Padre San Ignacio de los Ríos Yaqui y Mayo—with headquarters at Tórim. The rectorial system was eventually applied to all Jesuit missions in Sinaloa and Sonora. By the end of the seventeenth century four more had been established in Sonora, each of which comprised several *partidos*, or districts.[9]

The large population of the Mayo and Yaqui Indians within their respective deltas formed compactly settled areas that were easily missionized within a short time. North of the river deltas, however, the various native language groups were dispersed along the narrow valleys of La Serrana; this pattern resulted in a slow, often sporadic, and piecemeal evangeliza-

tion. As early as 1610 groups of Pima Bajo, or Nébomes, living along the
Río Tecoripa, a tributary of the Yaqui north of the delta, had visited Jesuit
missions in Sinaloa and had asked for resident priests to tend their people.
As they were engaged in winning over the Mayo and the Yaqui, however,
the few available missionaries were unable to oblige the request until 1619,
when they began baptisms at three Pima villages—Cumuripa, Suaqui, and
Tecoripa, the last serving as *cabecera* of the valley. Two years later Pima
Bajos, along the middle course of the Yaqui River, received priests at Onavas and Tónichi and at Movas and Nuri along the Río Chico, a Yaqui
tributary. Seven years elapsed before Jesuits moved into mixed Pima
Bajo–Eudeve country to the north and west, where the mission *cabecera*
of Mátape and its *visitas*, Nácori and Alamos, were established.[10] Late in
the seventeenth century Mátape was to become the main educational center (with the founding of the Colegio de San José) and one of the richest
missions in Sonora.

Later in the seventeenth century the Jesuits first formed missions among
the Opata of eastern Sonora. For several years el Gran Sisibotari, chief of
the Opata of the Sahuaripa Valley, had been asking missionaries to enter
his country. Like some of the Pima Bajo leaders, he had visited the missions in Sinaloa and was impressed by the church schools and agricultural
innovations that the Jesuits had introduced. Finally, in 1627, Father Pedro
Méndez, who had been working among the Mayo and the Yaqui, founded
missions in the large Opata villages of Sahuaripa and Arívechi and at Bacanora in the valley immediately to the west. Drained by a northward-
flowing tributary of the Yaqui, the valley of Sahuaripa with its broad alluvial floodplain, was one of the most fertile in Opata country. Méndez
marveled at the gardenlike farm plots of the natives: "There are streams of
fine water which the Indians employ with no little ingenuity for irrigating
their fields."[11] He praised his charges for their temperance, their lack of
idolatry, and their bravery as warriors. In fact, throughout the colonial
period the Opata and their linguistic cousins, the Eudeve, were held in
high esteem by both the Spanish religious and secular factions; they were
considered the most tractable and dependable of all the aboriginal groups
of Sonora.

Back in the mission of Mátape the Jesuits made plans to evangelize the
Opata along the Río Oposura (since the early nineteenth century called
the Río Moctezuma). In 1629 Father Martín Azpilcueta began baptisms in
the twin villages of Batuc (Santa María and San Javier) near the confluence
of the river with the Río Yaqui; in both towns missions were established.[12]
But because of insufficient manpower, difficulties in communications, and

the need to consolidate gains already made, expansion of the mission frontier stopped for nearly a decade.

Movement northward was revived in the late 1630s with the evangelization of the well-populated Sonora River Valley, the upper portion of which was one of the strongholds of the Opata, and through which Spanish exploring expeditions had passed a century earlier. In 1636 priests from Mátape began baptisms among the Pima Bajo at Ures on the middle Río Sonora, and from that base, three years later, missions had been made upriver among the Opata at Aconchi and Huepac; the neighboring hamlets of Banámichi and Baviácora served as *visitas*.

During the 1640s Jesuit missionizing in Sonora was once more partially arrested, this time by a dispute with the Franciscan order, whose priests from New Mexico had begun to penetrate the northeastern portion of the province. The Franciscan intrusion into nominally Jesuit territory was instigated by Captain Pedro de Perea, the founder of the first secular settlements in Sonora.[13] In 1644 this ambitious officer, with the concurrence of the Spanish government, recruited a group of laymen and five Franciscan priests from New Mexico to settle the northern part of the province. For five years (1644–1649) the Franciscans sought converts among the Opatas in the Bavispe and upper Sonora river valleys, along the Río de Fronteras near the present United States–Mexico border, and among the Eudeve in the upper San Miguel Valley, where Perea had established his headquarters. Understandably alarmed by this intrusion, in 1650 the Jesuit superiors successfully arranged a peaceful solution with the Franciscans, whereby the latter would limit their missionary activities to an area east of the upper Bavispe River, which would exclude them from Sonoran territory.[14]

Even prior to the 1650 concordat, the Jesuits lost no time reclaiming areas that the Franciscans had visited. As early as 1645 Jesuits had taken over Opata villages in the upper and lower Bavispe valleys, and some years later established missions at Arizpe and Chínapa on the upper Sonora river, as well as at Cucurpe on the upper San Miguel. Later (1653) they were firmly in control along the Fronteras River, where the Opata town of Cuquiárachi was made *cabecera* of the *partido*, with *visitas* in three smaller villages. Located on the Opata frontier, this *partido* would later bear the brunt of savage Apache raids from the north.

Attempts to convert the Seri hunters and gatherers in the Sonoran Desert to the west met with little success; in 1679 a small number of desert dwellers were persuaded to gather at the mission *visita* of Pópulo de los Seris on the lower San Miguel River, where they were to become farmers, but few did.[15]

By the mid-seventeenth century Jesuit hegemony extended over most of eastern Sonora, or La Serrana, home of the relatively advanced Opata, Eudeve, and Pima Bajo Indians. To the north and west of Opata country (the Opatería) lay the lands of the Pima Alto on the frontier of Christendom. The evangelization of these rather intractable farmers did not occur until the late 1680s, when renowned Jesuit priest Padre Eusebio Francisco Kino for the first time extended church control northward into present-day United States territory and westward into the northern part of the Sonoran Desert.

Kino arrived in Mexico from Spain in 1681. Following an abortive attempt to missionize Baja California, he was sent, five years later, to Sonora to begin conversion of the Pima Alto in extreme northwestern Mexico. In 1687 Father Kino, accompanied by two Jesuit priests, began baptizing at the village of Cosari, some ten miles north of the Río San Miguel mission of Cucurpe. At Cosari he established the mission of Dolores, which became his base of operations for the evangelization of all the Pimería Alta of northern Sonora and southern Arizona.[16] His efforts and those of his fellow churchmen would represent the last phase of Jesuit missionizing in northwest Mexico.

A few years after the first baptisms at Dolores, the Jesuits had visited and made plans to construct churches at San Ignacio and Caborca along the Magdalena River; along the Altar River at the *cabeceras* of Tubutama and Saric; and along the well populated Río Santa Cruz, where Suamca, Guévavi, and San Javier del Bac were made the head missions.[17] San Javier del Bac, with *visitas* near Tucson, Arizona, formed the northernmost of the Jesuit missions in Sonora. Finally, Kino established a small outpost at the Pápago seasonal camp of Sonoíta on the spring-fed stream of that name, perhaps to serve more as a way station on the trail to California than as a true religious center in a desolate desert.[18] The missions of Sonoíta and Caborca and Caborca's *visitas* along the lower Magdalena River were products of the Jesuits' sole attempts to penetrate the desert of western Sonora. Although during the eighteenth century missionaries had made exploratory trips beyond the Gila River of Arizona, because of insufficient personnel and danger of Apache raids, no missions were established in that territory.[19]

By the end of the seventeenth century the Jesuits had founded some thirty-eight head missions and fifty-nine *visitas* in Sonora (including those of the Yaqui and Mayo). Economically, the missions were for the most part self-sufficient, especially in terms of foodstuffs for both priests and natives. To obtain church materials, such as ornaments and vestments, and other items thought necessary for the proper discharge of their religious duties, the priests relied on the annual supply train from Mexico City. Once each year the padres would list their requests (*memorias*), which were relayed to the Jesuit steward (*procurador*) in the capital. He purchased the items and sent them on their way to Sonora via the mule caravan (see Appendix A). The funds for the purchases came from the royal treasury (Patronato Real), which by Spanish law allotted each mission *cabecera* an annual stipend of three hundred pesos. Because the freight charges usually amounted to half the stipend, the missionaries were forced to utilize other resources to supply their needs. This involved the sale of surplus agricultural goods to outsiders—mainly Spanish miners, who paid in silver, which was then sent to Mexico City to supplement the royal allowance.[20]

Under Spanish law a mission was to be dissolved ten years after its founding. The edict was based on the supposition that after that time Indians should have been sufficiently versed in Christian rules of conduct so they could be safely granted secularization. But under Jesuit practice mission status in Sonora was prolonged to the extent that secularization did not occur until the order was expelled in 1767. And even then most of the missions were taken over by the Franciscans, whose control in some cases lasted beyond the end of the century.

The underlying reason for the Jesuits' desire to continue mission status indefinitely was their belief that the Indians were not yet prepared to compete with Spanish lay society, which had been penetrating Sonora since the mid-seventeenth century. Moreover, as a secularized mission, or parish, the inhabitants living therein would be subject to payment of church tithes, and the priest in charge would no longer be eligible to receive the annual stipend from the royal treasury.[21] Understandably, the Spanish lay settlers, interested in acquiring church lands and cheap Indian labor, favored secularization. Jesuit priests and lay officials were thus at odds over this question, resulting in acrid recriminations from both sides, beginning especially in the early 1700s.[22]

Effects of the Church on Indian Life

There can be no doubt that the Jesuit mission system in Sonora was responsible for various degrees of culture change among the native population. In several instances, as with the Yaqui, the Mayo, and even the Opata, conversion to Christianity did not completely destroy native religious beliefs; in a sense it added to the local pantheon of deities and their functions and led to a harmonious blending of the two religions.[23] This phenomenon occurred widely throughout much of the Americas evangelized by various ecclesiastical orders. One of the important Jesuit rules for the spiritual education of the Indians involved special concern for children, whose minds could be molded more easily than those of adults. In each *cabecera* seminaries were to be formed in which children were to be taught to "read and write so they can serve the Church and give a wholly virtuous example to the rest of the pueblos."[24] Just how common and effective such instruction became in the mission system is unclear; teaching in both Spanish and the native language may have been limited to catechism and prayer.[25] Many of the missionaries formed church choirs among their neophytes, who were reported to be excellent singers; and some priests introduced European musical instruments, which the more talented Indians adopted readily to accompany the singers.[26] (One wonders if the small, often disharmonious, village bands that today are so common in rural Mexico may not be a heritage of colonial church influence.) Although the Jesuits did not intentionally encourage all Indians to learn Spanish, most of the Opata and many of the Yaqui were speaking it by the mid-eighteenth century.[27] Exposure to the language, however, may have been more through lay settlers than through the missionaries.

Probably the most enduring cultural change that the missions made on Indian life was in agriculture, the economic mainstay of most of the Sonoran natives. Through the introduction of Old World crops, farm tools and techniques, and domesticated animals, the church in part revolutionized farming and often increased or stabilized food production among the Indian farmers.

After religious conversion and settlement consolidation, one of the first agricultural tasks of the priests involved the establishment of a bipartite system of farmland division. A portion of the cultivable land, usually one or more plots of variable size that could be readily irrigated, was set aside as church land, or *labores comunales*, to be tilled collectively by the local inhabitants. The produce therefrom was for use by the priests and, in time of need, for the sustenance of the mission Indians; surplus could be sold to outsiders, the proceeds used for the upkeep of the local church. All other cultivable land was retained by the natives, to be farmed as they wished for their own needs. During the planting and harvest periods able-bodied males in the mission settlement were required to work the church lands three days weekly, their own plots three days, with Sundays and feast days reserved for compulsory church attendance and religious celebrations.[28]

The attitude of the Indians toward these rules is unclear, for the reports of the priests and other documents fail to dwell on the subject. It appears, however, that the natives accepted the system without rancor. One reason for their acquiescence may have been that they were paid for their labor by receiving daily at least two (and often three) substantial meals during the work period. According to Father Segesser (1737), who served among the Pimas, breakfast consisted of *atole*, a maize gruel flavored with fruit or chile, and the meal after the day's work was a large portion of *pozole*, a rich soupy mixture of hominy, chile peppers, and large chunks of beef, all the ingredients taken from the church storeroom. Father Segesser infers that the workers were also given tobacco imported from Mexico City and states that his Pima men "will work all day for a pipeful."[29] In addition, workers and their families periodically received from the padres clothing and cloth imported from central Mexico; such gifts may have softened the natives' attitude toward "forced" labor in the church fields. Father Pfefferkorn states that the food and income derived from those fields was such that "without them not a single mission could have survived."[30]

Little is known about the size of the communal fields during the early years of the missions; the padres may well have taken the best village cropland. In 1735 two of the three communal fields at the mission of Sahuaripa in eastern Sonora were approximately 9.5 and 13.5 acres (4 and 5.5 hectares), respectively (see Appendix B). In 1790, well after the Jesuit expulsion, Franciscan reports indicate that church lands of various pueblos along the lower Bavispe valley ranged from 50 acres (20 hectares) to 16 acres (6.4 hectares). The large size of these fields probably was not typical of missions of an earlier period (see Appendix C).

As introducers of European crops and farming techniques, the Jesuits actually served as agricultural extension workers; the mission was the center for the diffusion of agricultural innovations among the natives. Other religious orders, especially the Franciscans and the Augustinians, played a similar role in various parts of Mexico. In the small gardens that were usually attached to the church building or to the priest's house, the padres invariably experimented with many kinds of European fruits and vegetables. For example, Father Segesser at his mission in Tecoripa in Pima Bajo country, planted lemon, orange, fig, pomegranate, pear, and apple trees, as well as grape vines. Among his vegetables were chickpeas (garbanzos), lentils, peas, cabbage, garlic, and onions.[31] The Indians became interested particularly in lentils and chickpeas, which they added to their native legumes of pinto and tepary beans; cultivated cabbage, garlic, and onions often supplemented the native potherbs, such as the amaranths, gathered in the wild. Old World fruits introduced early into Sonora included watermelon, cantaloupe, and other melons, all of which the Indians readily adopted, growing them in sandy soils along streams. Sugarcane was also grown in mission fields; its cultivation was gradually adopted by the neophytes, who boiled down the juice into a syrup that, for sweetening, supplemented native honey gathered in the wild.

To cultivate his mission garden the priest must have had European tools—hoes, spades, rakes—the use of which he taught his Indian helpers. In their own plots the natives probably continued to use their dibble, or planting stick, to cultivate maize and beans until late in the colonial period.

Wheat was by far the most important field crop that the Jesuits introduced into Sonora. In its homeland—the Near East and the Mediterranean—wheat is a winter crop, planted during the cool-season rains, harvested during the summer dry season. The plant does poorly in areas of summer rain, when rust diseases invade. Thus, in most of Mexico wheat is cultivated during the dry winter season with the aid of irrigation. In northwestern Mexico the Jesuits discovered that the crop did poorly in most of Sinaloa; but as they advanced northward into Sonora they entered an area of dual-season precipitation, most of the rain falling during the hot summer, a lesser amount during the winter months. Taking advantage of the winter *equipata* rains, the padres, aided by irrigation, were able to produce substantial wheat harvests in the narrow valleys of La Serrana. The fact that most Spaniards, including the Jesuit priests, preferred wheat bread over the native maize foods was the basis for the introduction and expansion of wheat culture in colonial Mexico.

Although the mission Indians in Sonora continued to cultivate the native maize, beans, and squash on their own plots, the priests taught them to grow wheat, one of the principal crops produced on church land. With wheat was introduced the main tool for its cultivation—the wooden plow, or ard, drawn by a team of oxen. Seed was broadcast on the newly plowed soil, then covered by raking a crude brush harrow over the surface. When needed, canals were constructed to conduct river water to the field for irrigation. To harvest the grain Indians were taught to use the iron sickle, and threshing was accomplished on the *era*, a hard-surfaced spot, by driving mules over the sheaves. Finally, the grain was separated from the straw and the chaff by winnowing.

During the early years of the missions, the padres employed Indian women to grind wheat on native *metates*, or stone querns, but eventually mule-powered grist mills of stone (*tahonas*) were introduced to produce the flour for making the bread that the padres, like most Europeans,[32] so dearly cherished as their staff of life (fig. 12). Loaves of bread were baked

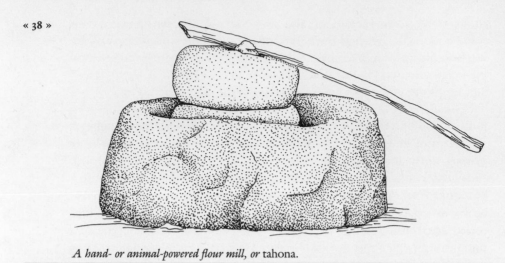

A hand- or animal-powered flour mill, or tahona.

FIG. 12

in a domed-shape outside oven, an ancient Old World device still used in many parts of Mexico.

Gradually, Indians became accustomed to cultivating wheat and other small grains, such as barley, on their own lands and to using flour for making tortillas. In Sonora to this day thin flour tortillas are the norm in the rural diet, rather than those of maize common elsewhere in Mexico. To cultivate wheat the natives were perforce obliged to make use of the wooden plow. By the eighteenth century they were making their own crude plows, which they eventually used to prepare the ground for planting maize. Father Pfefferkorn, a Jesuit of German background, who worked among the Pima Alto in the mid-eighteenth century, observed that "the Sonoran plow is so constructed that a European can hardly understand how a field can be worked with it."[33]

The introduction of domesticated animals and the business of stock raising into Sonora was perhaps less significant than the new crops in changing the lifeways of the native inhabitants. Initially, the large animals—cattle, horses, mules—or *ganado mayor*, frightened the natives and often invaded their fields, destroying crops; the Indians were more comfortable with the smaller, more docile animals—sheep and goats—or *ganado menor*, which they were able to control. For the Spaniards, including the missionaries, however, livestock was a part of their way of life.

The Jesuit economic system included the raising of livestock on a comparatively large scale. At least one ranch, or *estancia*, was established for each mission *partido*. Located some distance from the mission settlements to avoid crop damage by livestock, a ranch was usually managed by a *mayordomo*, often a mixed-blood mulatto or a mestizo well versed in herding techniques; later in the eighteenth century the care of mission livestock passed largely into the hands of Indians.[34]

The animals multiplied rapidly in the semiarid bush and grass cover of the Sinaloan coastal lowlands and in the lower mountain slopes of La Serrana in eastern Sonora. Mesquite and other leguminous plants combined with various grass species to afford abundant browse and graze for both *ganado mayor* and *menor*. However, a serious plague in the form of wild predators—cougars, jaguars, wolves, and the ubiquitous coyotes—often reduced the annual crop of calves, foals, and lambs.[35]

The livestock population belonging to the missions is difficult to estimate, mainly because of the varying numbers recorded in occasional reports made by local priests for the visitor general of the Jesuit order. In a series of reports made in 1690, the number of cattle in the *estancias* of given *partidos* ranged from around 300 to 1,000 or 2,000, of horses and mules, somewhat less, and of sheep and goats, from 200 to more than 1,000. The *partido* of Cucurpe on the upper San Miguel River had two *estancias*, one at Saracachi in the hills east of the river with 1,500 head of livestock, the other at Meresichi, north of the mission, with only 230 head. Another report of the same date indicates that the *partido* of Oposura on the Río Moctezuma had 1,400 cattle, 1,100 sheep, 500 horses and mules.[36]

Probably all of these figures represent undercounts. The actual number of livestock on the range could never be counted with any accuracy, because many of the cattle had become feral, or half wild, a result of inadequate herding practices that permitted the animals to roam and reproduce freely in the bush. In some instances greater care was taken with horses and mules, both valued for packing and riding. Sonoran mules, especially, were in demand for transportation of merchandise. For breeding, one stallion or one jack was provided for twenty-five mares; such a group was termed a *manada*. On each *estancia* annual or semiannual roundups, or rodeos, were made to break in horses and mules, brand calves and foals, and to butcher cattle for the supply of tallow, hides, and meat for the mission centers.[37] In all of the *cabeceras* the Indians were permitted to celebrate every Catholic feast day by consuming the fresh meat of one or two beeves. But the missionaries also oversaw the preparation of dried and salted meat (*tasajo*) for future consumption and for sale to outsiders.

According to church documents, from 1650 to 1680 the cattle herds in *estancias* surrounding Mátape mission increased from 600 to an incredible 50,000 head, making it possible for the Jesuit fathers to supply seed

« 40 »   livestock for the missions farther north. In the 1680s plans were drawn to make annual cattle drives of 5,000 head from Mátape to the Jesuit-operated sugar mills in central Mexico. At least one drive occurred, via the coastal lowlands of Sinaloa, but a herd of only 1,750 head reached its destination near Puebla, east of Mexico City.[38] By the 1690s livestock in the Mátape area had been reduced to 12,000 head,[39] due possibly, in part, to deterioration of pasture through overgrazing. Today the hills around Mátape, covered with bush and cactus, could hardly support even a few hundred cows, much less the herds of the size of those in colonial times.

On the mission *estancias* in mountainous eastern Sonora sheep and goats were nearly as numerous as cattle and horses. As were the larger animals, sheep were permitted to roam the range without much care, resulting in poor quality wool and serious losses in the annual lamb crop.[40]

In the Yaqui pueblos, however, sheep were of greater importance than cattle or horses. There, the Indians took readily to sheep raising. In cleared areas along the riverbanks grass grew abundantly in the fertile alluvium. By the mid-seventeenth century some families had begun to obtain their own flocks, which they tended close to the pueblos. They constructed brush corrals to protect the animals overnight, to keep them from the planted fields, and to care for the ewes during lambing season.[41] The wool clip was sold to lay Spaniards in mines and ranches in eastern Sonora, apparently without protest from the Jesuit priests. A census document dated 1698 reports a total of 12,350 sheep and goats, including both Indian and church flocks, in six of the eight Yaqui villages within the river delta, in contrast to only 2,956 head of *ganado mayor*.[42] Nearly a century later, in four of the pueblos the proportion of sheep and goats (21,416) to cattle and horses (2,133) was even greater.[43] The Yaqui sold raw wool to Spanish merchants, and the women wove woolen blankets and cloth with their native horizontal ground looms (fig. 13). In 1774 a Spaniard introduced the Yaqui to the European hand-and-foot loom, the use of which greatly increased the output of woolen goods in the pueblos.[44] In the mission pueblos of northeastern Sonora the Opata women also were adept at weaving cloth, but mainly of cotton, rather than of wool, with a native loom similar to that of the Yaqui.[45]

Although some of the priests kept household animals, such as pigs and poultry, the native Sonorans were slow to adopt them, despite the presence of the domesticated native turkey in some pueblos. Father Segesser had some two hundred chickens, and both eggs and meat were consumed on festive occasions, but he complained that even his personal Indian servants would habitually steal his eggs rather than raise their own hens.[46]

An aboriginal Yaqui ground loom.

FIG. 13

Eventually, some Indians acquired burros for transport, since, as Father Nentvig observed, these animals "could live on little, even on garbage piles around the pueblos."[47] Milk cows may have been kept only by Jesuits of German extraction, since Spaniards and Indians rarely drank milk; many rural Sonoran villagers still drink little milk. Oxen used for plowing church lands were mission property, and were probably tethered in or near the villages during planting time.[48]

Throughout eastern Sonora, the Jesuits logically retained the aboriginal practice of maintaining settlement on the natural terraces that overlook the river flood-plains. Among both the Opata and the Pima Bajo, the larger native villages were usually chosen as the church center of a given mission district, or *partido*, and Indians living in scattered hamlets along the rivers were urged to congregate in the larger pueblos. With little interest in town planning, the priests of colonial times permitted the natives to retain an amorphous arrangement of houses surrounding the mission; the grid street pattern with a central square, or plaza, today so common even

Arizpe, Sonora River Valley, La Serrana, eastern Sonora. Built on the Pleistocene terrace, the town overlooks the river floodplain in the middle distance.

FIG. 14

in small villages of Mexico, probably was not introduced into the mission towns of Sonora until after the colonial period. Moreover, the construction of substantial churches of stone or adobe, even in the larger villages, was usually not undertaken until well after the first baptisms had taken place. Again, the padres, being satisfied with native dwellings, appear to have had little influence on house construction within their missions. The aboriginal Opata and some Pima knew adobe construction techniques, but the few Jesuit accounts of native life in Sonora that comment on Indian house types describe beehive-shaped dwellings and rectangular houses with walls of wattle or woven mats sometimes mud-daubed, and with flat or slightly sloping roofs of mats covered with dirt. The substantial flat-roofed or pitched-roof adobe structures that predominate in most of the larger towns of eastern Sonora today probably date from postcolonial times. Today, along the river valleys of La Serrana the rural villages, with the church rising above the squat one-story dwellings, in part represent a legacy of the mission period of eastern Sonora (figs. 14 and 15).

Bacanora, La Serrana, eastern Sonora.

FIG. 15

Sonora represented one of the most successful mission endeavors of the Jesuit order in the New World, possibly on a par with its organization in Paraguay, established about the same time. But in the eyes of many Spanish secular administrators, the Jesuits were too successful economically and politically, which led to their expulsion from the American colonies in 1767 by royal edict.

CHAPTER 4

Spanish Settlement of Sonora: The Mines and Ranches

Spanish laymen, mainly miners and stockmen, began to enter eastern Sonora about the time that Jesuits were founding missions among the Opata. By the end of the seventeenth century the mining of silver ores had become firmly established in Sonora, with the development of several sizable settlements (*reales de minas*) and a multitude of smaller ones (*realejos*). None, however, attained the size and opulence of Zacatecas, Taxco, or Pachuca in central Mexico. Spanish miners were often stockmen as well, with herds concentrated in the mountain scrub and grasslands mainly in northeastern Sonora.

Mineralization along fault planes and at contact zones of plutonic intrusions with sedimentary rocks occurs abundantly in northwestern Mexico, producing silver, gold, and copper ores. The semiarid climate of the region, which prevents excess leaching and erosion, has likely contributed to a normal concentration of rich surface ores along outcrops. This surface enrichment facilitated mining by Spaniards,[1] who had only superficial knowledge of the primitive methods of ore extraction and reduction in vogue during the seventeenth century. By digging pits, or "gopher holes," along a vein outcrop, miners could obtain fairly rich ores. But the zone of surface enrichment was usually shallow, and as ore quality decreased with depth, most mines soon became unprofitable and were abandoned. Moreover, Spanish miners of the time were unable to cope with flooding when they attempted to excavate below the water table. Thus, most of the early mining camps in Sonora were ephemeral affairs and few became stable or permanent settlements.

Even before the Jesuits missionized the area, eastern Sonora may have been visited by Spanish prospectors who crossed the Sierra Madre Occidental from the northern plateau in search of silver and gold deposits. A document concerning a civil court case in 1673 indicates that two of the witnesses, both Spanish miners, claimed they had been in Sonora since 1633, "before the province was conquered and the Indians Christianized."[2]

Historians have credited Captain Pedro de Perea, commandant of the garrison at Villa de Sinaloa, as the first colonizer of Spanish settlement in Sonora, probably in 1641.[3] Perea may have begun operations in Sonora several years earlier. In 1634 he led an expedition against the Opata living along the Bavispe River in the northeasternmost part of the province.[4] To accomplish that, he and his soldiers would have had to trace the entire length of Sonora, from Sinaloa to the present United States–Mexico border, thus gaining an acquaintance with the valleys and mountains of La Serrana. In 1637 he was authorized by the viceroy of New Spain in Mexico City to colonize Sonora (called Nueva Andalucía by Perea) and to promote the development of mining and agriculture there at his own expense. The viceregal authorization also gave the captain temporary use of twenty-five soldiers from the Sinaloan garrison to help carry out his plans.[5] Documents indicate that as early as 1637 Perea's soldiers had entered Sonora, reaching the upper San Miguel Valley, where they established for their captain a ranch called Nombre de Dios near the Eudeve village of Tuape. In reply to a letter dated March 21, 1640, from the Jesuit visitor general, Padre Leonardo Jatino, Pedro de Oliva, a lieutenant in Captain Perea's expedition and in command of the troop detachment at Nombre de Dios, indicated that he and the soldiers had been in Sonora for nearly three years, or since 1637. (Oliva's reply was dated March 24, 1640.)[6] Oliva wrote further that

> the soldiers of don Pedro de Perea have been in this province serving His Majesty, hoping to increase the royal income by searching for mines. In this they have succeded, thanks be to God, by the discovery last month of four mines [ore deposits] near the Valley of Sonora, two leagues east of the village of Sinoquipe. The ore from these deposits has been taken to the Real de los Vírgenes [in the Sierra Madre Occidental] for assaying, because at present there is no one in this province who understands well the reduction of the ores, but through our own attempts at smelting we are certain that they are of silver.[7]

« 46 »   The discovery of 1640 appears to have been the first of its kind in Sonora, and would lead to more finds, which began a profitable mining industry in the province.

Apparently, at that time Pedro de Perea was still in Sinaloa and may not have joined his troops until 1641. In that year he journeyed to Parral, the de facto capital of Nueva Vizcaya, where he obtained authority from the governor to continue his colonization scheme under the title of magistrate (*alcalde mayor*) of Sonora. Because most of his soldiers already in Sonora were obliged to return to their garrison in Sinaloa, Perea looked to New Mexico for more settlers, who, together with five Franciscan missionaries, arrived in 1644.[8] Eight of the dozen or more lay settlers from New Mexico were members of the Pérez Granillo clan.[9] Francisco Pérez Granillo, the clan head, had been with Perea's soldiers in the San Miguel Valley since 1637,[10] and at least four of the garrison troop members who came in that year stayed on to become important Sonoran miners.[11]

After Perea's death in 1645, the search for silver mines continued, and a few years later the Spaniards had established the *reales de minas* of Santiago de los Reyes, San Pedro de los Reyes (both in the mountains between the San Miguel and Sonora valleys), and Nacatóbari (of uncertain location but probably in the mountains east of the Moctezuma River near the present-day mining camp of Lampazos) (fig. 16). In 1649 Santiago de los Reyes housed the headquarters of the *alcalde mayor* of Sonora (Simón Lazo de la Vega). At that time the Spanish lay population in the entire province included about forty-five men, according to a roster of militia formed to quell an uprising of Pima Alto north of the Sonora Valley.[12] Nearly all of the Spaniards were miners, most of whom had originally come from Parral; at least one was a rancher who offered to donate one hundred head of cattle and three hundred *arrobas* (seventy-five hundred pounds) of dried meat to feed the militia and 880 Opata allies on their military expedition. Probably few Spanish women accompanied their men to Sonora at that early date. The widow of Pedro de Perea, Doña María Ibarra, however, lived with her younger children in the Opata village of Banámichi in the Sonora Valley.[13]

Consolidation of the Mining Industry

About 1657 rich silver ores were discovered in the mountains between the Sonora and Oposura (Moctezuma) valleys, a few leagues southwest of the Opata village of Cumpas. Called San Juan Bautista, this *real* became one of the foremost producers of silver in Sonora during the mid-seventeenth

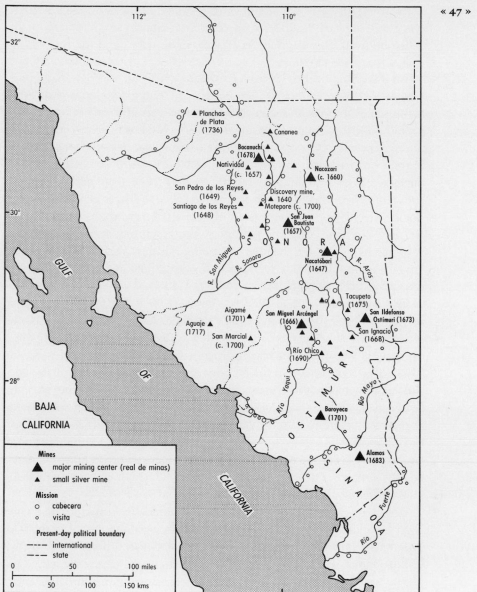

Mining centers and missions, Sonora, seventeenth and early eighteenth centuries.

FIG. 16

« 48 »   century and served as the administrative capital of the province for nearly one hundred years. Because of ore depletion and flooding of mine shafts by groundwater, production decreased by the 1680s, and the *real* was finally abandoned in 1750.[14]

An indication of the richness of surface deposits at San Juan Bautista is seen in the amount of silver registered at the *real* in 1659, two years after its founding. In that year some fifty miners reported a total of 8,050 *marcos*, or about 4,050 pounds, of refined silver produced from ore taken from the shallow diggings on a high hill immediately south of the settlement.[15] By 1681 output had greatly decreased; only four deep and badly flooded mines were in operation, seven small smelters sufficed to reduce the ores, and sixteen stores cared for the commercial needs of the local inhabitants.[16] Some thirty-five years later most of the inhabitants of the Sonoran capital and its vicinity were ranchers; mining had practically ceased and all but two merchants had left the town for better business elsewhere.[17] San Juan Bautista exemplifies the sudden rise and gradual fall typical of most mining centers of colonial Sonora.

The early mines of San Pedro de los Reyes and Nacatóbari continued to yield silver beyond the 1660s, but that decade saw several new discoveries that led to an increase in both silver output and Spanish population in Sonora. The mines of Nacozari came into production about 1660.[18] Located in the oak-covered hills bordering the upper reaches of the Río Moctezuma, Nacozari became a major producer of silver during the late seventeenth and early eighteenth centuries.[19] In the 1750s it was finally abandoned, more because of dangers from Apache raids than because of ore depletion.[20] Beneath the surface deposits of silver lay vast quantities of copper ores that were exploited by North American interests in the nineteenth century. The same copper mines still produce under Mexican management.

About two decades after the Nacozari mines were discovered, many silver ore deposits were being mined in the mountains adjacent to the Río Bacanuchi, a northern tributary of the Sonora River. These new finds led to the establishment of the Real de Bacanuchi in an area already settled by several stock ranchers. Silver production continued well into the eighteenth century, and in the 1760s new ore discoveries were made.[21] Located on the northern frontier, the Bacanuchi Valley and its adjacent mines were continually threatened by Apache assaults, resulting in its abandonment at the end of the century.

Farther south, in 1666 new silver lodes were uncovered in the highly mineralized mountains a few leagues west of the Yaqui River and probably

near the present mining center of San Javier.[22] San Miguel Arcángel, the   « 49 »
real de minas that developed amidst a multitude of small diggings, for a
time was as productive as Nacozari. One of the best mines belonged to
one Doña Juana de Jaziola (possibly a widow who inherited her husband's
estate). Between 1677 and 1680 her mine, located in La Barranca, some
distance from the town center, yielded annually nearly one thousand *marcos* (eight thousand ounces) of silver.[23] The ore was reduced by means
of amalgamation, one of the few instances of the use of mercury in processing silver ores in colonial Sonora, where smelting was the principal
method employed. Mining operations at San Miguel continued until the
mid-eighteenth century.[24]

By the last quarter of the seventeenth century, with the establishment
of new mining centers and growth in stock raising, the Spanish population
had increased perhaps four-fold over that suggested by the militia roster
of 1649. According to similar rosters taken in 1684, 186 men of Spanish
descent responded to a call to arms for protection against an impending
Indian uprising on the northern frontier. The militiamen, miners, and
ranchers, were distributed as follows:[25]

Real de San Juan Bautista	36
Valle de Teuricachi	12
(near present United States–Mexico border)	
Valle de Tepache	29
(stock-raising area)	
Real de Nacatóbari	11
Real de Bacanuchi	62
Real de San Miguel Arcángel	36

Two years after the discoveries at San Miguel Arcángel, another significant find occurred in the area known as Ostímuri, a mountainous section
between the Mayo and Yaqui rivers. There, in 1668, a small *real de minas*,
San Ignacio, or Los Gentiles, evolved near a series of ore outcrops.[26] Later,
in 1673, San Ignacio was eclipsed by the founding of a much-larger *real*
nearby, San Ildefonso de Ostímuri. Three years later the civil province of
Ostímuri was formed, with the residence of the *alcalde mayor* at San
Ildefonso.[27]

The new province, or *alcaldía mayor*, included the territory between the
Mayo and the Yaqui rivers and extended northward to the Río Aros, taking in the Sahuaripa Valley and its many villages. Late in the seventeenth
century numerous small mining centers arose in the mountainous terrain
of Ostímuri, but after 1700 most, including San Ildefonso, began to de-

cline, owing to a decrease in ore quality and threats of rebellion of the Tarahumar Indians in the Sierra Madre to the east.[28] In the southern part of the province, however, the late 1600s and early 1700s saw the rise of several profitable mines, such as Río Chico (ca. 1690) and Baroyeca (1701). In 1720 the capital of Ostímuri was moved to Río Chico, but the magistrate often stayed in the larger Baroyeca,[29] probably to monitor better the registration of silver bullion and the availability of Indian labor.

In the province of Sonora few silver mines were opened in the eighteenth century. Those of Motepore near the Río Sonora Valley, just west of Sinoquipe mission, were discovered around 1700, and twenty years later the new *real* had become a sizable Spanish settlement of two hundred miners, stockmen, and their workers; by mid-century it had been virtually abandoned because of Apache raids.[30] An early attempt on the part of Spanish miners to penetrate the Sonoran Desert is seen in the establishment of two *reales* based on silver and gold deposits well south of the lower Río Sonora: Aigamé in 1701 and Aguaje sometime before 1717. The former was small and its mines were worked sporadically during most of the century, but it was strongly revived in the 1770s and again in the 1790s.[31] The *real* of Aguaje, the larger of the two, by 1717 contained ten stores that supplied food and equipment to five mines; the ores were reduced in three large smelters.[32]

One of the oddest mining discoveries in Sonora occurred in 1736, when a Yaqui Indian found by chance several large, partially buried ingots of nearly pure silver at a place called Arizona, or Planchas de Plata, not far from the present United States–Mexico border, some forty kilometers southwest of Nogales. Each of the larger pieces weighed more than three *arrobas* (twenty-eight kilograms), but one was estimated at twenty-one *arrobas* (over one hundred kilograms); the entire trove amounted to some 2.5 tons. After lengthy court proceedings the Spanish Crown claimed the treasure, as it decided that the pieces were products of processed ore, which probably was mined from deposits in the immediate vicinity of the find.[33]

Mining Methods

Mining and ore-reduction processes that the Spaniards used in Sonora were far more primitive than those practiced in the large *reales* of central Mexico. Reference has already been made to the shallow open-pit diggings ("gopher holes") commonly excavated to reach the enriched oxidized surface ores along a vein outcrop. Later, vertical shafts (*pozos*) were sunk one or two fathoms (two to four meters) into the weathered material and small horizontal drifts may have been extended along the vein. Official ordinances promulgated to regulate mining practices in New Spain appear to have been little observed in Sonora, far from central authority, despite occasional inspections made by the provincial *alcalde mayor*.[34] Only for the larger *reales*, such as San Juan Bautista, is there documentary evidence of the construction of deep shafts, drifts, and stopes along rich veins and the use of wooden beams and cribbing to reinforce hanging walls.[35] But there, as elsewhere, once the water table was reached at depths of thirty to fifty meters, flooding usually decreased production and eventually led to mine closure. At San Juan Bautista the miners made various attempts in the late seventeenth century to control flooding by forcing Indian workers to carry out water in rawhide bags (contrary to labor laws). The workers negotiated the exit by climbing up crude chicken ladders—notched logs—which was how they hauled ore to the surface.[36] The use of water lifts, such as the *malacate*, and the construction of drainage addits, common in mines on the central plateau, seem to have had no place in Sonora.

Mining tools are occasionally mentioned in inventories that accompany documents involving civil suits. For example, the assets of one Simón García, a miner living near the *real* of Tacupeto, Ostímuri province (1707) included iron crowbars of twenty-five pounds weight, augers, picks, sledge hammers, and a forge for repairing tools.[37] Although known in the mines of central Mexico early in the eighteenth century, the practice of blasting with black powder went unmentioned in the early colonial documents regarding Sonora, where it may not have been commonly used until the latter part of the century.[38]

Ore-reduction methods in Sonora may have been as primitive as mining techniques. Silver was extracted by first crushing the ore to gravel size either by hand with hammers or occasionally in small, mule-powered stamp mills.[39] This operation was followed by smelting the crushed ore in crude adobe or stone furnaces (*hornos castellanos*) that employed lead or litharge (either lead oxide or lead carbonate) as reagents and charcoal for fuel. Lead ores abounded in Sonora, and the ingredient was exported to mining centers on the plateau, such as Cusihuiriáchic, north of Parral.[40] Air blast for the furnace was supplied by hand-operated sheepskin bellows equipped with copper nozzles, which, with the iron implements (tongs, rakes, pincers, and so forth), all had to be imported from Parral or Guadalajara at great expense.[41] For making charcoal, wood, especially mesquite and oak, was plentiful; but around Sonoran mines the special profession of *carbonero*, or charcoal maker, did not develop, as it did in the

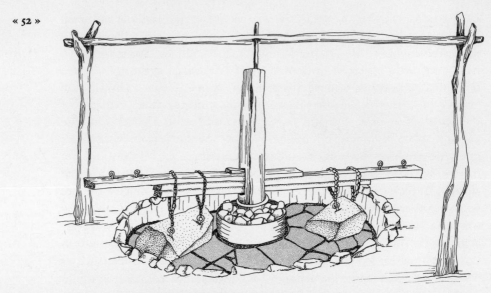

An arrastre, *or animal-powered mill for pulverizing ore.*

FIG. 17

large *reales* of central Mexico; Sonoran miners owned and operated their own kilns with Indian laborers whom they supplied with steel axes, also imported.[42]

Owing mainly to high price, uncertain supply, and exorbitant freight costs for mercury, the amalgamation process was little used in Sonora. Where it was employed, along with smelting, as at Nacatóbari, San Miguel Arcángel, and Alamos, ore was comminuted to a fine powder in the *arrastre*, an Old World invention consisting of a circular sunken, stone-floored platform in which two large boulders attached to a sweep were dragged by mules over the ore (*arrastrar* means to drag) (fig. 17). The use of the *arrastre* became much more important late in the eighteenth century, when mercury became more available.[43] Besides mercury, two other reagents were required in the amalgamation process: salt and magistral (a copper sulfate). Salt came from the coastal salines near the mouths of the Yaqui, Mayo, and Fuerte rivers,[44] and copper sulfate ores were plentiful in many parts of Sonora and Sinaloa.

Silver bullion produced by either smelting or amalgamation was in the form of platelike pieces of many sizes and weights, called *planchas, tejos,* or *tejuelos.* These were exported via merchants to central Mexico and eventually were resmelted into bars destined for Mexico City or for Spain. Smaller bits of silver, or *tepusques*, were used by Sonoran miners as a medium of exchange to purchase food and other necessities from merchants, Indians, and Jesuit priests.[45]

Despite the relatively primitive mining methods, the last half of the seventeenth century was the period of major silver production in the provinces of Sonora and Ostímuri. Captain Juan Mateo Mange, *alcalde mayor* of Sonora (1701–1703), estimated that in the previous fifty years, thirty million pesos of silver had been produced in his province alone, and during the two years of his tenure as *alcalde*, silver bullion worth half a million pesos was registered through his office.[46]

During the seventeenth century most of the silver produced in Sonora and Ostímuri was hauled by mule train across the Sierra Madre to Parral, where it was transshipped by oxcart to Durango for assessment of the royal severance tax. According to manifests housed in the Parral archives, nearly 40 percent of the silver shipped out of Parral to Durango during the period 1670–1688 originated in Sonora and Ostímuri.[47] Using documents in the Archivo General de Indias, Seville, Spanish historian Luis Navarro García estimated that during approximately the same period the average annual production of silver in Sonora amounted to a third of that mined in all of Nueva Vizcaya (northwestern Mexico).[48]

The Real *of Alamos*

One of the most productive and enduring mining centers of northwestern Mexico was Alamos, today in the southernmost part of the state of Sonora, but during colonial times within Sinaloa Province. Rich silver deposits in the vicinity of the town were discovered in 1683 and exploited sporadically for more than two hundred years, until the Mexican Revolution of 1910 caused the permanent closure of most operations. Alamos was the only *real de minas* in present-day Sonora that was perhaps comparable to the big mining centers of central Mexico.

All of the mines were located in the Sierra de Alamos southwest of the town (fig. 18). Most of the exploitable deposits were found in mineralized veins along the eastern side of an intrusive porphyry dike at its contact with the sedimentary and granitic country rock of the sierra.[49] As in most of the vein outcrops in colonial mines, the surface ores were highly oxidized and enriched to a depth of 150 to 180 meters, easily mined, and amenable to seventeenth-century methods of ore reduction—smelting or amalgamation, depending on the chemical nature of the ores.[50]

The initial discovery of silver in the Sierra de Alamos may have been made in 1683 at a place called Promontorios near the crest of the range.[51]

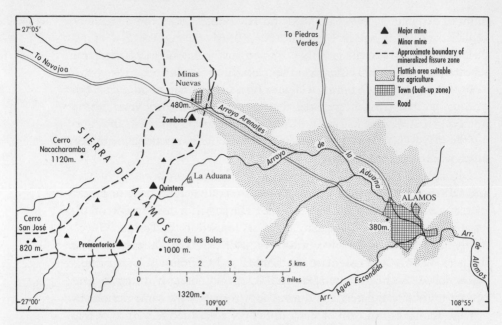

The Alamos mining district, southern Sonora.

FIG. 18

A veritable silver rush ensued; six years later Father Kino, who passed through the area on his way to Sonora, commented that forty-three rich "diggings" (probably shallow open-pit cuts along vein outcrops) were being operated in the vicinity.[52] Early on, most of the miners were living at the *puestos* (places lacking official sanction as villages or towns) called La Aduana at the northern base of the sierra and Los Alamos a short distance farther east. Although government officials thought neither site suitable for the establishment of a *real de minas* (due mainly to inadequate water supply), Alamos eventually became the main habitation and commercial center of the area.[53] Situated at the eastern end of a wide, arable basin with mean annual rainfall of nearly five hundred millimeters (twenty inches), the town and adjacent mines could depend on a substantial local supply of basic foodstuffs.

Alamos grew rapidly toward the end of the seventeenth century. In addition to the Promontorios mines, in 1709 a new discovery, La Quintera, was made farther down the northern slope of the sierra near La Aduana; and about 1750 a third large mine, Zambona, was opened at the base of the sierra near the newly founded settlement of Minas Nuevas.[54]

During the mid-eighteenth century La Quintera was the major silver

producer in the Alamos area, despite the inundation of some shafts and
galleries by groundwater.[55] At that time Alamos was known as one of most
productive mining centers of New Spain and, with a population of some
three thousand, as the largest commercial market in northwestern Mexico.
La Aduana, however, contained most of the silver beneficiating plants.[56]
In 1776 nearly two-thirds of the silver produced in Sonora and northern
Sinaloa came from the Alamos district, which at that time contained some
of the richest mines of New Spain; less than 15 percent came from small
mines in eastern Sonora.[57]

To feed the urban folk and the Indian laborers in the mines, wheat
flour, maize, and beans were purchased from the Jesuit missions in the
Yaqui and Mayo deltas nearby, supplementing the local agricultural production in the Alamos basin.[58]

Although present-day Alamos maintains a pleasant colonial ambiance,
practically all of the old mining centers of Sonora have long been abandoned and few traces remain of their former existence. As on the site of
San Juan Bautista, today only scattered slag heaps, a few stone foundations
of adobe structures, shards of imported polychrome pottery from central
Mexico, pieces of wine jars, and bits of Chinese porcelain are indicative of
small towns that at one time had a church, ten to twenty stores, and several
score inhabitants. Such sites are ripe for field surveys in historical archaeology. In eastern Sonora the ephemerality of the colonial mining centers
in the mountains contrasts strongly with the permanence of the mission
villages in the adjacent river valleys.

Merchants and Miners

Merchants and their mule trains played a significant role in the commercial
relations of the Sonoran mining centers and the outside. Most merchants
who operated in Sonora were based in Parral, where they had access to
goods and middlemen from central Mexico. They supplied the miners
with the iron tools necessary for extracting ore and the equipment used in
smelting and amalgamation (for the latter, especially, the government-
controlled supply of mercury, whenever available). They brought in luxury
items, like wine, brandy, olive oil, and tobacco. Craftware, mainly pottery,
leather goods, hats, and so on, from central Mexico also was imported.
But the principal import was cloth of many types: coarse cottons and
woolens manufactured in the small mills (*obrajes*) of Puebla and Queré-
taro, luxury silks and ribbons from Spain and the Orient, as well as tai-
lored clothing, including footwear, for male and female alike. Cheap cloth

« 56 »   was in great demand in the mines; in the absence of coins it was the main form of wages paid to Indian workers. Concerning this practice a document of 1671 stated: "Ropa es la moneda que corre entre los indios" (Cloth is the money that circulates among the Indians).[59]

The more affluent miners, especially those serving as local officials, usually obtained necessities from merchants through consignments. But a familiar part of any *real de minas* was the small shops (*tiendas* or *pulperías*) that catered to the less-prominent miners and prospectors and to Indian workers. Some of the traveling merchants set up stores in the *reales*, but most were probably run by local citizens. In one case, an *alcalde mayor* of Sonora Province, Francisco Cuervo y Valdés, began his term of office in 1681 by importing from Parral thirty thousand pesos' worth of merchandise, mostly cloth, to sell in his own stores in Nacozari, Bacanuchi, and San Miguel Arcángel.[60] In general, the size and importance of a *real de minas* at any one time could be judged by the number of stores: in 1681 San Juan Bautista boasted sixteen, San Miguel Arcángel twelve, Nacozari, three.[61]

Because they controlled the supply of merchandise and usually had ready cash, many merchants and store owners became the "bankers" in a mining center. In exercising this function they were able to grubstake poor prospectors and miners and to make loans at exorbitant interest rates or to participate in the proceeds of a lucky mining discovery.

Merchants also performed another important function: they hauled out silver bullion produced in a given *real de minas*. As indicated earlier, during the seventeenth century most of the bullion of Sonora was packed by mule train across the Sierra Madre to Parral, where it was assayed and shipped to Durango and Mexico City. Some miners may have carried out their own bullion, but most was consigned to traveling merchants who were prepared to make a profitable return trip to their home base in Parral. To carry out this function, official manifests indicating the weight of the silver in *marcos* were issued to merchants by local authorities and signed by the owner of the metal.[62]

The terrain and mule trails that linked Sonora and Parral across the sierra were difficult and dangerous. In the northeastern part of the province, three of the easiest, but least direct, roads became the most dangerous after the beginning of the Apache incursions in the late seventeenth century (fig. 19). Without military escort all three were practically untenable. The northernmost actually rounded the sierra via Guadalupe Pass along the present United States–Mexico border. The trails through Púlpito and Carretas passes farther south were easily negotiable and were

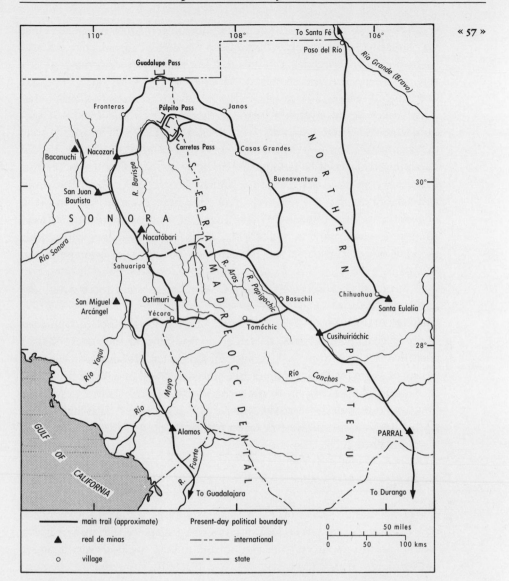

Main trails joining Sonora and Chihuahua, seventeenth and eighteenth centuries.

FIG. 19

often termed "the main routes between Sonora and Parral," having been used as early as the 1650s by muleteers.[63] But via either road, once on the plateau the mule trains were still some 80 leagues (450 kilometers) north of Parral, necessitating a long, arduous journey through semidesert country and often besieged by warring Indians.[64]

To the south lay the most direct but difficult trails to the plateau. One, via Sahuaripa, was supposed to have been blazed by Pedro de Perea in 1641 on his journey from Sinaloa to Parral,[65] but there is no evidence that this trail was used to any extent during the colonial period or since. It passed over rugged ranges and through difficult canyons to reach the Papigochic River Valley on the plateau, whence mule travel on to the mines of Cusihuiriáchic and Parral would not have been difficult. A much easier mountain trail passed through the Pima pueblos of Yécora and Maicoba and the Tarahumar villages of Yepáchic and Tomóchic to reach Cusihuiriáchic and Parral. By 1779 this trail was made an official post road between Chihuahua and Sonora and many merchants may have used it earlier.[66] Today it is negotiable by jeep, the eastern half being an improved paved highway. Another trans-sierran trail lay far to the south, connecting the plateau with the lowlands of Sinaloa via the Topia mines, but it was of little use to the Sonorans.

A more usable but lengthy road followed the coastal plain through Sinaloa and Nayarit to Guadalajara on the central plateau and on to Mexico City. Taken occasionally by merchants bound for Sonora throughout most of the colonial period,[67] this trail became important for the export of gold and silver bullion from Sonora and Sinaloa after a branch of the royal treasury (a *caja real*) was set up in Alamos in 1769 and moved farther south to Rosario in 1781.[68] By mule train the journey from Alamos to Mexico City took six months, and because of the many rivers that had to be forded, the coastal trail was traversed mainly during the dry season, from October through May.[69]

Spanish Livestock Economy

Though not as profitable or widespread as mining, stock raising early became an occupation and lifeway for many Spanish lay settlers in eastern Sonora. In the seventeenth century several miners owned herds of cattle, but few could compete with the well-established Jesuit *estancias*. As early as the 1640s, one of the first Spanish cattle ranches was formed by Pedro de Perea Ibarra, son of Captain Perea, the colonizer of Sonora. The son chose an ideal site for his endeavor—the lush mesquite and oak grasslands of the Bacanuchi Valley in the northern part of the province, at the time far from any Jesuit mission.[70] Seed stock for the early ranches may have been driven in from Chihuahua, but later most Spaniards probably purchased cattle, horses, and mules from the Jesuits. By 1685 there were six livestock ranches in the Bacanuchi Valley, four in the valley of Teuricache, near the present United States–Mexico border, one near Opodepe in the San Miguel Valley, another near the Río Sonora, and nine in the mountains around the *real* of San Miguel Arcángel.[71]

Although some stockmen claimed to have obtained large ranches (*sitios*) through royal grants (*mercedes*),[72] most miners probably had no legal title to the land on which their animals fed; they practiced an open range system. Early in the eighteenth century several Spaniards had established cattle ranches and haciendas in the foothills adjacent to the Jesuit missions in the river valleys throughout eastern Sonora. In 1713, for example, one José de Zubiate, a miner from the *real* of Nacozari, grazed seven thousand head of cattle near Oposura mission in the Moctezuma Valley, much to the consternation of the Jesuits, who claimed that the animals were invading the cultivated fields of their Indians.[73] At that time the padres also alleged that six Spaniards were running at least twelve thousand head in the mountain scrub between the Sonora and the Moctezuma valleys and in the Llano de Tepache east of Oposura.[74] A few years later (1723), eighteen Spanish cattlemen in or near the San Miguel Valley were being inspected in regard to cattle brands and land titles, which indicates increasing government control over the livestock industry in the province.[75] In time, the Spanish miners-stockmen and their mestizo and mulatto vaqueros would take over the best pastures of eastern Sonora, a process that was accelerated especially after the Jesuit expulsion in 1767.

In contrast to the scarcity of livestock around the Parral mines on the northern plateau of Mexico (because of frequent raids by Toboso nomads), during the 1660s cattle and mules in Sonora and Sinaloa were described as being so abundant that they sold for little ("donde por la abundancia casi no tienen precio").[76] Hence, after fulfilling their needs for animal products (dried meat, hides for ore sacks, tallow for candles, and the like), the miners with large herds often drove excess animals across the Sierra Madre to Parral and other centers on the plateau.[77] Such drives continued into the next century, for in 1715 a miner, José de Zubiate of Nacozari, drove six thousand head of his livestock from the Llano de Tepache to the mines in Chihuahua.[78] Even the Jesuits may have engaged in such drives. One report recalled that when the mission at Mátape possessed large herds in its surrounding *estancias*, as many as three thousand head of cattle were driven to Chihuahua, where they sold for as little as three or four pesos per head.[79]

CHAPTER 5

Mine and Mission Relations in Colonial Sonora

There was probably no other area of colonial Spanish America in which mines and missions were in such close proximity as in Sonora. This phenomenon led to an economic and social relationship that was mutually opportune but often antagonistic.

Lacking the means to produce their own food, most miners relied on imports of grains, meat, and other animal products to sustain themselves and their labor force. The most convenient source for such necessities proved to be the nearby missions, which usually harvested surplus crops from their communal fields and which on their *estancias* bred livestock in excess of their immediate needs. Miners also required cheap labor to excavate and process ores and to perform other tasks in the *reales*. Again, the mission Indians presented the logical source, a source legally sanctioned under the Spanish *repartimiento* labor system.

For their part, the Jesuits usually welcomed the opportunity to supply food to the miners in return for pieces of silver, which the priests used to help purchase church furnishings and luxury items, as well as cloth, clothing, and such things as tobacco, which helped maintain the loyalty of their neophytes.[1] For many Spanish secular officials, however, such transactions led, rightly or wrongly, to the perception that the priests were accumulating undue wealth, contrary to church precepts.[2] On the other hand, the Jesuits naturally opposed the miners' legal right to the biweekly levy of Indian labor for the mines and strongly objected to the miners' frequent abuse of that right. Such action resulted in almost continuous disputes between the missionaries and the lay society in Sonora for much of the colonial period.

Archival documents dealing with some of the disputes reveal interesting information regarding the role of the missions in supplying food for the mines. During the 1660s and the 1670s Spanish officials made various charges against Padre Daniel Angelo Marras, visitor general of the Sonoran missions and rector of the Jesuit college at Mátape, concerning the practice of selling food to the miners.[3] To answer the charges, Padre Marras assembled a group of miners from the *real* of San Miguel Arcángel and other mining centers to testify in his behalf. Hearings were held in the Sonoran capital of San Juan Bautista in 1671 and in San Miguel Arcángel in 1673. Since its establishment in 1666 San Miguel had depended on the missions of Tecoripa and Mátape for food. At the 1673 hearing twenty-one witnesses, all miners, indicated that San Miguel could never have survived as a *real de minas* and that the entire mining industry of Sonora would have collapsed without the food supply from the various Jesuit missions. Several witnesses stated that wheat flour was the principal item that miners purchased from the missions, because Spaniards could not cope with maize foods such as tortillas or atole. (But maize was also purchased to feed Indian workers.) One witness also averred that during the eight years from 1663 to 1671 the missions had distributed some 150 *cargas* of wheat flour (1 *carga* equals 4 *fanegas*, or 6.4 bushels) to the Sonoran mines (mainly to San Juan Bautista and San Miguel), and that because of growing Spanish population in the province and increasing demand for flour, the padres were prompted to increase production by installing mule-powered grist mills (*tahonas*), rather than relying on the old method of hand-grinding on the native metates. Said another witness: "Before [installation of] the *tahonas* there was hardly enough bread available to eat with a serving of chocolate."[4]

Other agricultural products that missions furnished miners included beans and other legumes, sugarcane and molasses (*chancaca*), and hides, tallow, and candles.[5] In 1695 Mátape mission alone dispensed the following to mines in various parts of Sonora at fixed prices: eighty *cargas* of wheat flour at twenty pesos/*carga*; three *cargas* of molasses at fifty pesos/*carga*; two *cargas* of soap at sixty pesos/*carga*; thirty *arrobas* of tallow at six pesos/*arroba*; thirty *arrobas* of lard at six pesos/*arroba*; and seventy *arrobas* of meat (dried) at two pesos/*arroba*.[6]

According to some detractors, the Jesuits established a virtual monopoly on the sale of agricultural products by charging the miners such

« 62 » low prices that Spanish merchants and stockmen were nearly always undersold.[7] For example, the missions charged three pesos for one *fanega* of maize, whereas merchants demanded six to ten pesos.[8]

In at least one documented case, the relationship between missionaries and miners in connection with food supply resulted in a singular activity on the part of the priests. Mining in Sonora was not limited entirely to lay Spaniards. The Jesuits of Mátape mission also engaged in the extraction and refining of silver ores from deposits near Tecoripa, not far from San Miguel Arcángel. According to the rules of the Jesuit order, priests were forbidden to own, operate, or even acquire knowledge of mining.[9] But apparently, ownership of the mines in question had been signed over to the mission by a Spanish miner in payment for debts he incurred for supplies obtained from the padres. Moreover, the priests claimed that the mines belonged to the College of San José at Mátape, not to the mission itself, and thus the superiors permitted continuation of the "forbidden" activity.[10] During the late seventeenth century annual proceeds from the Tecoripa mines under church management ranged from three thousand to twelve thousand pesos, a substantial windfall for the college and the missionaries.[11]

Although the Jesuits discouraged commercial transactions between their Indians and the Spaniards, many Indians, especially the Opata and the Yaqui, commonly sold maize to miners in exchange for pieces of cloth.[12] According to one account of 1673, "It is public knowledge that in the Valley of Oposura [Moctezuma] each year the [Opata] Indians barter more than two thousand *fanegas* [about thirty-two hundred bushels] of maize with the miners and merchants."[13] But because most Sonoran Indians were subsistence farmers, they had little interest in raising crops in their own fields to sell to outsiders, an attitude that the padres encouraged. During the latter part of the colonial period, Yaqui Indians were noted for their commercial activity and carried maize, salt, and other items long distances with their own mule trains to sell or barter in Spanish communities. The Yaquis, however, always failed to add freight expenses to the asking price, thus underselling most Spanish merchants, who thought the practice ridiculous and unfair.[14]

Labor Supply for the Mines

The labor supply proved to be one of the most vexing problems in the relationship between the Jesuits and the Spanish secular authorities in colonial Sonora. In Mexico the type of labor system that the conquerors imposed on the native people tended to vary with the cultural status of the aborigines. In central Mexico, inhabited by sedentary farmers, the large mining centers depended mainly on forced native labor recruited through the *repartimiento* system, whereby a given number of males from surrounding villages were required to work in the mines for a number of weekly or biweekly periods during the year at a minimal wage. In north-central Mexico, inhabited by belligerent nomads, a forced labor system (except for occasional enslavement) was unworkable; there, free native labor, composed largely of immigrants from central Mexico, became the rule. In northwestern Mexico, however, the presence of the tractable Cáhita, Pima Bajo, and Opata sedentary farmers afforded the possibility of imposing the *repartimiento* system to supply the mines with needed labor. In some cases, free voluntary labor was also available, particularly from the Yaqui Indians, some of whom became even more adept than Spaniards as prospectors.

The *repartimiento* developed gradually in New Spain during the sixteenth century. The system was codified in 1609; thereafter it underwent various modifications and persisted in places until the end of the eighteenth century.[15] Thus, by the time mining developed in Sonora after 1640, the rules of the system had been well established. By law only 4 percent of the male population in a given pueblo could be taken to work in the mines at any one time, and these could not be detained for an excessive period. After completing the work period each Indian was to be paid by the mine owner for both the labor and the time used in travel to and from the village. The assignment of the work forces was to be determined and supervised by a *justicia ordinario*, or an *alcalde mayor* of the district or province involved.[16]

In Sonora, Sinaloa, and elsewhere in Mexico, Indians who made up a *repartimiento* work gang were called *tapisques*. A *sello* was a work order or document issued by a secular official in behalf of a miner or rancher who had applied for Indian labor. This document indicated the number of workers to be called up from a given village; it was presented to the native governor of the village by a *topile*, usually a trusted Indian who would be responsible for conducting the *tapisques* to their destination. Such *sellos* were often kept among the papers of the Jesuit priest in charge of the village from which the workers were taken, and a few of these documents are extant today in archival collections. (See Appendix D for an English translation of late-seventeenth-century examples found in the Archivo Histórico de Hacienda, Mexico City.)

The *repartimiento* system appears to have operated fairly smoothly dur-

in various mines and ranches throughout Sonora, especially during the eighteenth century. Having witnessed the application of the *repartimiento* system in Sonora during the late seventeenth century, Padre Kino wisely obtained from the Spanish government a provision that for twenty years excluded mission Indians in the Pimería Alta from forced labor in the mines or on Spanish haciendas.[20]

With the gradual decline of forced labor in eastern Sonora in the eighteenth century, the number of free Indian workers increased, and by the end of the century they made up the prevailing type of laborers in the mines. Spearheading that trend in Sonora and Sinaloa, Yaqui men, as early as the mid-seventeenth century, were leaving their homeland in the Yaqui River delta to seek temporary employment in some of the newly founded *reales de minas*.[21]

A major cause of this phenomenon initially grew out of occasional crop failures in the delta, which forced Indians to seek subsistence elsewhere, despite the efforts of the missionaries to keep them in the pueblos.[22] Some Mayo and Pima Bajo also joined the free labor force, but in smaller numbers than the Yaqui. Eventually, many natives became skilled free workers in the mines, operating furnaces in the smelting process or tending stamp mills, while the *tapisques* performed less demanding tasks, such as ore extraction and comminution by hand methods. The Yaqui wandered throughout Sonora and even migrated into the northern plateau of Mexico, seeking employment in the mines of Parral and elsewhere in Nueva Vizcaya.[23]

Despite their total lack of knowledge of mining and metallurgy before Spanish conquest, some Yaqui, Mayo, and Pima Bajo became expert prospectors and led Spaniards to productive silver, gold, and lead deposits. In 1774 even the viceroy in Mexico City opined that "the Yaqui and Mayo Indians are and have been of great utility to the public treasury through the knowledge that they have in prospecting and discovering veins [of minerals] . . . and thus they should not be subject to tributes that at present have not been assessed in their native land [the province of Sonora]."[24]

Although Spanish miners needed workers on a continual basis throughout the year, the Sonoran Indians, both free and *tapisques*, established the practice of returning to their villages for the "*temporada*," a period of at least three months during the summer rains to plant crops and attend native ceremonies. Thus, productive mining was limited to only eight or nine months of the year, a practice encouraged, of course, by the Jesuits but strongly opposed by the lay Spaniards.[25]

Just as the Jesuit missionaries changed Indian lifeways in various ways, so the miners, merchants, and ranchers in Spanish lay settlements modified native race and culture in Sonora. The influences that the Indians—both free workers and *tapisques*—were exposed to in the mines were more negative than positive, especially in the eyes of the padres. Indian workers were introduced to Spanish customs and language through free association with riffraff, or the *vagabundo* lower class—mulattoes, mestizos, and other mixed bloods—common in all colonial Mexican mining towns. According to the Jesuits, miscegenation was rampant in the mining centers. Indian males cohabited with mixed-blood women, and Indian women, who often accompanied their native husbands, were subjected to temptation by both Spaniards and mixed-blood males, who offered colored cloth and dresses for sexual favors.[26] Moreover, in order to accumulate a dependable labor force, miners tried to persuade *tapisques*, with offers of plentiful food, cloth, and "freedom from the persecution of the padres," to remain in the mines.[27]

Such contacts led, especially among the Opata and the Eudeve, to loss of tribal cohesiveness and eventually to the loss of language. Most of the Pima Bajo followed the same path. In 1900 (Mexico 1900) only 30 Opata and some 250 Pima speakers were recorded in the official census for Sonora, and today both languages can be considered extinct.[28] In contrast, despite their wanderings and mining associations, both the Yaqui and the Mayo even today have maintained their language and many of their native customs. Edward Spicer has discussed reasons for the tenacity of Yaqui culture in northwestern Mexico;[29] a similar study is necessary to the understanding of the Mayo people.

As a more positive influence Spanish officials claimed that many Indian workers in the mines learned valuable trades, such as carpentry, tanning, cobbling, and tailoring,[30] but the same might be said of similar opportunities in the missions. In any case, Hispanicization of the Indians was probably more thorough and rapid in Spanish centers than in the missions.

Another process that helped change (and disrupt) Indian lifeways was the gradual incursion of Spanish stockmen and farmers into mission village lands in the river valleys of eastern Sonora. Despite vigorous Jesuit protests, this illegal usurpation began in the late seventeenth century, continued apace for half of the eighteenth, and increased in intensity after the

« 68 » Jesuit expulsion of 1767, when the missions were placed under temporary control of the Franciscan order. As early as 1722, many Spanish ranchers, farmers, and miners had penetrated the Valley of the Río Sonora, taking over mission lands. In a sixteen-league (sixty-kilometer) stretch of the middle portion of the valley between the Real of Motepore (near the pueblo of Banámichi) southward to the Real of Concepción (near the present town of Mazocahui) some twenty-five Spaniards and their mixed-blood hirelings had established fourteen cattle ranches, eleven farms, and eight smelters amidst the four mission pueblo lands of Banámichi, Aconchi, Huepac and Baviácora.[31] Similar incursions were taking place in the valleys of the Río San Miguel and Río Oposura (Moctezuma).[32]

By 1767, within the upper and middle section of the Río Sonora Valley, the proportion of Spaniards and mixed bloods (*gente de razón*) had grown to 40 percent of the total population.[33] Again, in 1799, the population of the upper San Miguel Valley was approximately half Indian (Eudeve) and half Spanish.[34] The census of 1780 indicated that in all of the province of Sonora 37 percent of the population was composed of Spaniards and mixed bloods.[35] Cynthia Radding (1990) has detailed the emergence of a peasant class among Indian and mixed ethnic groups in eastern Sonora during the late colonial and early independence periods, emphasizing patterns of social change and cultural persistence.

In Sonoran mission, mine, and stock ranch alike, European diseases took their toll of the Indian population, as in most of the Americas. In some parts of Sonora near the end of the colonial period, Indians had been reduced to less than a tenth of their estimated pre-Conquest numbers. For example, in 1765 the mission population of the Opata and the Eudeve numbered only fifty-eight hundred[36] compared to a conservative estimate of sixty thousand prior to Spanish occupation. Likely, the mission policy of assembling Indians from scattered hamlets into larger pueblos aided the rapid spread of contagious diseases. The same may be said of the mining centers, where Indian workers were forced to live and work in close quarters.

Daniel Reff has documented the various epidemics that drastically reduced the native population of Sonora and other provinces of northwestern Mexico during the colonial period. According to Reff, the disastrous scourges of smallpox, typhus, measles, and other Old World diseases that swept through central Mexico soon after the Spanish Conquest may not have reached Sonora; but as early as the 1530s and the 1540s, plagues had decimated Indian villages along the Pacific lowlands of Mexico as far north as Culiacán in central Sinaloa.[37] By 1592, smallpox had reached what is now

southern Sonora, causing death and famine among the Yaqui and Mayo.[38]
During the first fifty years of Jesuit missionization, epidemics are said to
have killed half of the Mayo population.[39] For much of the seventeenth
century, the Jesuit priests reported several epidemics among their mission
Indians in central and eastern Sonora,[40] but the great scourge of *matlaza-
huatl* (probably typhus) that plagued central Mexico from 1736 to 1739
apparently did not reach Sonora, although it was reported in Baja California.[41] Much of the eighteenth century saw epidemics appear every ten or
twelve years in Sonora and other areas of northwestern Mexico,[42] and according to Padre Pfefferkorn, a contagion of smallpox that ravaged southern and central Mexico entered Sonora in 1765, "leaving not a single person
unaffected," and few Indian adults survived.[43] The same epidemic was reported from Arizpe: "This province has been very sick because of smallpox
and *metlasaque* [spotted fever?], from which many have died, and even
some of the padres became ill because they have had to preside over so
many burials, but none have died."[44] The decimation of native population
through disease was as manifestly severe in Sonora as in the rest of Mexico
and Central America during the colonial period.

CHAPTER 6

Indian Depredations in Sonora

Among the most serious calamities to befall Sonora in the eighteenth and nineteenth centuries were the raids on mines, missions, and stock ranches inflicted by nomadic tribes—the Apache, Jocome, Jano, and Suma from the arid north and the Seri and, occasionally, the Pima Alto from the desert west. Major attacks by the Apache began in the 1680s and continued sporadically for two hundred years, until the last of the bands led by the renowned chief, Gerónimo, was finally defeated by the combined action of Mexican and United States troops in the 1880s.[1] By the early 1700s Seri bands had advanced into the western frontier of Spanish settlement in Sonora, and in the 1750s they were joined by neighboring Pima Alto Indians, who devastated much of the northwestern part of the province. Both tribes were finally subdued by the end of the century (fig. 21). Even the usually trustworthy Yaqui rose against the Spaniards briefly (1740–1742), spreading fear and destruction in south-central Sonora (Ostímuri Province).

By the mid-eighteenth century scores of mines and ranches in northern and eastern Sonora had been abandoned, commerce came to a standstill, and most of the province endured a long period of economic depression from which it did not recover until late in the nineteenth century. Spanish settlements, owing largely to their smallness and dispersed distribution in mountainous areas, fared much worse than the larger river-valley mission towns, which because of their size and compactness could muster some protection against the forays of the nomads. Among the mining centers only the Alamos area in the south escaped the raids. In 1770, with a population of some seven thousand, it contained half of all Spaniards in Sonora, three thousand in the *real* of Alamos alone.[2] A poignant reminder of how far and how late the Apache penetrated is seen in an old cemetery in the village of Nuri, four hundred kilometers south of the northern frontier. There, a tombstone reads "On the morning of the twenty-eighth day of October, 1878 Don Victorino Ramonet, aged 52, was killed by the Apaches in the Arroyo Hondo, 8 leagues from Nuri. His wife and son, consumed by the most profound grief, leave this marker in his memory" (my translation).

Causes and Methods of Indian Raids

Various reasons for the incursion of nomads into the Spanish areas of Sonora have been suggested, ranging from the effect on game and useful plants of prolonged droughts in the desert to the example set by the Pueblo Indians in their 1680 rebellion against Spanish rule in New Mexico.[3] There is substantial evidence for the drought factor. Although cited in colonial documents of the time as the major cause of the Apache invasion,[4] it is doubtful that the Pueblo uprising had much influence on the nomads, who had been engaged in small raids on Spanish ranches in northwestern Chihuahua since the late 1670s.[5] Probably the most logical explanation for the raids is simply the nomads' need of food, easily obtained by killing or stealing livestock on Spanish ranches. Such an endeavor was far easier than hunting wild game;[6] in the case of the Southern Apache, who had occupied the desert and steppe country in what is now southwestern New Mexico and southeastern Arizona, few bison or antelope herds were available for hunting, the main feeding range for those animals being far to the east in the high plains of eastern New Mexico and western Texas. Thus, the growing herds of horses, mules, and cattle on Spanish ranches were attractive sources of food.

According to opinions of the time, horse and mule meat became the favorite food of the Apache and other desert nomads on the northern and western Sonoran frontiers.[7] Jesuit priests Father Och and Father Pfefferkorn, both of whom worked among the Pima Alto, waxed eloquent in describing the Apaches' addiction to mule meat and their ways of preparing various parts of the animal for consumption. Both attributed what they considered to be the repugnant body odor of the Apache warrior to his feasting on mule meat, including the intestines encased in animal fat.[8] Horse and mule meat were always preferred over beef, although cattle were taken on raids to be skinned for hides. Sheep and goats were usually ignored. Both Apache and Seri gradually learned to ride horses, but they never bothered to learn horse breeding; they opted to raid Spanish settlements for their mounts, always choosing the best animals for riding.[9] One Spaniard observed that "they appreciate a good horse more than a wife or child."[10]

As mounted marauders the Apache could travel long distances quickly, cutting out animals from the Spanish and mission herds at will and driving them back to camps in southeastern Arizona for future use. During the late colonial period and well into the nineteenth century the Apache established regular trade in stolen goods, especially livestock, that Spaniards and Mexicans in New Mexico and Chihuahua were eager to buy.[11] The Indians took stolen horses and mules to sell at the livestock market in Santa Fé, where Sonoran brands were as common as those of New Mexico.[12] As late as the 1850s New Mexicans would visit Apache camps to purchase stolen animals,[13] and often ranches and mines in Chihuahua afforded an attractive market for Apache booty of Sonoran origin.[14]

For weapons the desert marauders, including Apache, Seri, and Pima Alto, employed the bow and arrow and occasionally the lance; the Apache probably did not obtain firearms until the mid-nineteenth century. Most raids occurred during moonlight, more often in winter than in summer, to take advantage of long nights for concealment; groups of thirty to sixty warriors would round up animals on a ranch, making off under cover of darkness.[15] By day the raiders hid in rugged mountain retreats to avoid Spanish cavalry. At any one time probably no more than four hundred Apache men were on various raiding parties throughout northern and eastern Sonora.[16]

Isolated mines and ranch headquarters were especially vulnerable to Indian attack, and when Spaniards or mixed bloods resisted they were usually killed; women and children were often taken prisoner to be sold as slaves or held for ransom at Apache camps.[17] In Spanish settlements houses of wattled walls and thatched roofs were easily set afire with flaming arrows, forcing occupants into the open.[18]

Destruction became so rampant in the mid-eighteenth century that government officials issued regulations to the effect that all houses occupied by Spaniards in Sonora were to be constructed of adobe bricks with flat roofs covered with sod or dirt to avoid destruction during Indian attacks.[19] Although occasionally assaulted, the Indians living in mission pueblos were usually successful in warding off attacks, especially the Opata, who were noted for their bravery in battle and their hatred for the

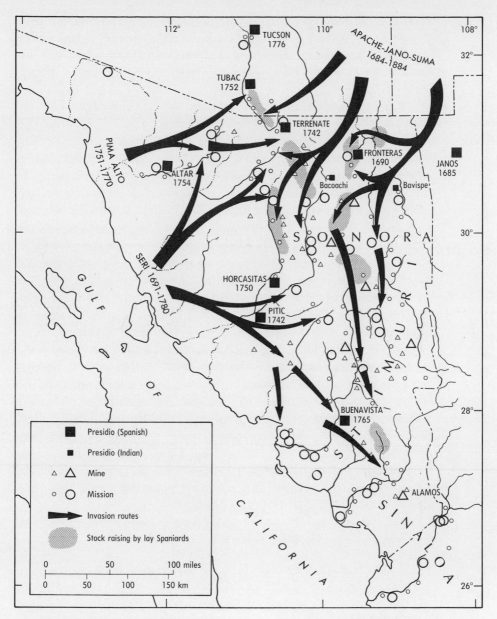

Incursions of hostile nomadic Indians into Sonora, late seventeenth through the nineteenth centuries. (Present-day political boundaries shown.)

FIG. 21

« 74 »   Apache. Moreover, the Opatas' use of poisoned arrows served as an effective deterrent. At least three Opata mission pueblos on the northeastern frontier—Baserac and its two *visitas*, Bavispe, and Huachinera—were actually walled to protect their inhabitants from Apache attacks.[20]

Effects of Indian Depredations in Sonora

From the time of the first major Apache raids on Spanish settlements along the northern frontier in 1684 until well into the nineteenth century, reports of local churchmen and secular officials form a litany of dire accounts describing abandonment of mines and ranches, loss of livestock, and murders of the citizenry in most parts of eastern and central Sonora.[21] In 1684 officials estimated that in a three-month period one hundred thousand head of livestock from ranches on the northeastern frontier had been lost to the nomads.[22] In following years the northern mining centers of Nacozari and Bacanuchi, with others, were partially deserted, in part because the Apache had driven off the mules used to haul ore, and mule trains that imported supplies and exported bullion had been attacked, paralyzing transport.[23]

With the rebellion of the Seri and the western Pima Alto, the first half of the eighteenth century saw a sharp increase in Indian depredations, resulting in further abandonment of Spanish settlement in northern and central Sonora. In 1737 Captain Juan Mateo Mange reported on the lamentable state of the province: "many mines have been deserted, 15 large *estancias* along the frontier have been totally destroyed, having lost two hundred thousand head of cattle, mules, and horses; several missions have been burned and two hundred Christians have lost their lives to the Apache enemy, who sustains himself only with bow and arrow, killing and stealing livestock. All this has left us in ruins."[24]

As late as 1852 the U.S. border commissioner, John Bartlett, informed the secretary of state in Washington:

> Much has been said of the depredations of the Indians in the State of Sonora, of the great extent to which they have been carried. These statements have not been exaggerated. None but those who have visited this State can form any adequate idea of the widespread devastation which has marked the inroads of the savage. Depopulated towns and villages, deserted haciendas and ranches, elegant and spacious churches falling to decay, neglected orchards teeming with fruit, and broad fields once highly cultivated, now overgrown with shrubbery and weeds show to what extent the

country had been overrun. . . . There is scarcely a family in the frontier
towns but has suffered the loss of one or more of its members or friends.
In some instances whole families have been cut off—the father murdered,
the mother and children carried off into captivity.[25]

Spanish Response to the Indian Menace

From the beginning of Indian troubles during the late seventeenth century, secular authorities in Sonora attempted to control the incursion of
nomads with military action. Reliance was initially placed on the formation of local militia, composed of a few mounted Spanish "professional"
soldiers, miners, and stockmen, usually accompanied by a large contingent
of friendly Indian "foot soldiers" from neighboring mission towns.[26] The
main function of these hastily and poorly organized expeditions was to
pursue and punish the bands responsible for committing atrocities and, if
possible to retrieve stolen livestock. Although militia continued to be employed even in the late colonial period, their effectiveness was minimal,
owing largely to poor-quality Spanish personnel, many of whom "had
never fired a musket in their lives."[27] If on occasion such campaigns were
successful, credit was due mainly to the Indian foot soldiers, especially the
Opata, who were far more adept at tracking and fighting than were inex-
perienced Spaniards.[28]

Aside from the failed attempts to quell Indian raids with local militia,
the Spanish government organized at least one effective military expedition against the marauders in the eighteenth century: the Expedición de
Sonora of 1767, aimed at destroying the Seri threat in central Sonora. It
was funded in part by contributions from the larger mining centers and
secular towns of Sonora and Sinaloa, from wealthy merchants in central
Mexico, and with supplies of wheat flour furnished by the Sonoran missions.[29] The expedition was manned by several hundred Spanish troops
and Indian auxilaries and temporarily halted the Seri menace at the battle
of Cerro Prieto in the desert of western Sonora.[30] Peace was short-lived,
however, for six years later the Seri were again on the warpath, penetrating
far eastward into Ostímuri, where they forced the abandonment of the
Real de Baroyeca and even threatened the large mining center of Alamos.[31]

The Spaniards' most ambitious attempt to stem the incursion of the
nomads involved the establishment of presidios, or forts, at strategic places
along the northern frontier. Since the time of the Reconquest in Spain,
the presidio had been an integral part of Spanish military strategy against

« 76 » enemy forces. The Spaniard looked upon the fortified stronghold, a product of Roman and medieval heritage, as a symbol of authority and strength, despite its frequent operational failure.

To counter early Apache thrusts into Sonora, presidios were established at Janos (1685) on the northern plateau in Chihuahua and at Fronteras (1690) on the river of that name in northern Opata country (fig. 21).[32] Staffed by a few poorly trained soldiers, neither presidio served its purpose, for the Apache warriors could easily enter Sonora by filtering through the Púlpito and Carretas passes between the two Spanish redoubts. The main activity of the military in both presidios proved to be escorting mule trains carrying cloth and other goods from Parral into Sonora and returning with silver bullion.[33]

Additional presidios were not founded along the frontier until the mid-eighteenth century, when Terrenate (1742) and Tubac (1752) were built on the Santa Cruz River near the present Arizona-Sonora border, and Altar (1754) was built on the river of that name, a tributary of the Magdalena in Pima Alto country. Nearly a quarter of a century later the last of the northern presidios was established at Tucson (1776), from which arose the present Arizona city on the Río Santa Cruz. The purpose of all of these later presidios was to protect Spanish settlement from Seri and western Pima Alto raids, a function poorly performed, for the nomads could plunder almost at will. Farther south the presidios of Pitic (1742, 1780), Horcasitas (1750), and Buenavista (1765) were all formed to stem Seri incursions.[34]

Perhaps the most effective type of fortified place along the northern frontier of Sonora was the "semi-presidio," composed of Opata Indians—one at Bavispe in the northeast, directly in the path of Apache forays, the other at Bacoachi on the upper Río Sonora. Still a third Indian company, made up of Pima Alto, was stationed near Tubac. All three were formed around 1780 as part of a general reorganization of the defense system on the northern frontier.[35]

The reorganization was instigated by the visitor general to New Spain, José de Gálvez, in 1765 and later by the *Instrucción* of viceroy Bernardo de Gálvez in 1786. Punitive expeditions were increased, trade alliances with some of the nomadic tribes were made, and many of the hostile bands were enticed, with Spanish gifts of food and horses—in a kind of peace-reserve system—to settle near the presidios. By 1790 the Spaniards and their marauding antagonists had achieved an uneasy truce that lasted until 1831. At that time the Mexican government withdrew the food handouts, and general warfare with the Apache resumed in Sonora and neighboring Chihuahua for most of the nineteenth century.[36]

According to William Griffen, the failure of the Spanish military and Mexican officials to make a lasting peace with the Apache may have stemmed in part from fundamental differences of sociopolitical thought between European and Indian. European ideas of hierarchy and subordination did not exist among the Apache, each band having temporary leaders who were not recognized by neighboring groups. Thus, from the Apache point of view a truce that Spaniards or Mexicans made with a particular group did not apply to members of other bands, who thought themselves free to plunder at will.[37]

During the latter part of the eighteenth century each presidio was normally manned by a contingent of fifty to seventy-five military personnel, but the troops were rarely maintained at full force. Officers, especially the captain, were often peninsular Spaniards, whereas the ethnic makeup of noncoms and troops ranged from *criollos* of Spanish blood to mixed-bloods of various categories.[38] Officers and inspectors frequently complained of the poor quality of the troops and their reluctance to face the enemy in battle.[39] In addition, every presidio had a group of ten to fifteen Indian trackers, or *exploradores*, to accompany the cavalry in pursuit of marauders. Many of the troops were accompanied by wives and children, and because many farmers and ranchers had fled with their families to the fort for protection, the total population of any one presidio often reached into the hundreds, leading to food shortages.[40] Normally, the supplying of one or more presidios with provisions was contracted out to merchants in neighboring towns.[41]

Following military norms of the day, presidios were laid out in quadrilateral plan, two hundred Spanish yards (*varas*) on the side, the barracks and other buildings facing inward around a center plaza or parade ground, the backs of the structures forming a rampart or outer wall.[42] Adobe walls and flat roofs characterized most of the buildings. Land around the fort served as pasture for the horses and mules numbering two hundred to three hundred head, used by the cavalry (*compañía volante*) and for packing on punitive expeditions. Unless well tethered and carefully tended, the animals were often fair game for any Indian enemies who happened by, and by stampeding the herd, the marauders could make off with a goodly number under the noses of the hapless soldiers.[43]

During the 1790s, when the Spanish government succeeded in settling various nomadic groups near the presidios, huts were constructed outside the garrison ramparts to form an incipient village. Some former presidios, such as Fronteras and Altar, eventually became towns, and Tucson and Pitic (Hermosillo) grew to be large cities. With the mingling of whites,

« 78 »   Indians, and mixed bloods, presidios, like the mining communities of Sonora, became centers of further miscegenation.

During the 1840s, when Apache raids were again in full swing, the Mexican government attempted to stem the attacks by establishing "military colonies," composed of citizens and foreigners from the central part of the country, at the sites of the old presidios of Bavispe, Fronteras, Santa Cruz (Terrenate), Tucson, and Altar. This measure, including frequent military campaigns against the Indians, proved to be ineffective, and the raids continued unabated.[44]

Sonora and Chihuahua were not the only provinces of northern Mexico to suffer the ravages of nomadic Indian depredations during the eighteenth and nineteenth centuries. Beginning in the 1760s Spanish settlements in Coahuila, Nuevo León, Texas, and Nuevo Santander (present-day Tamaulipas) bore the brunt of repeated attacks by the Comanches. These High Plains bison hunters of Shoshone stock had become the best riders among all the nomadic tribes of western North America.

Close parallels can be drawn between Sonora and the rest of northern Mexico immediately south of the Rio Grande regarding methods of plunder by the nomads and the Spanish response—ineffective defense despite construction of presidios, and resulting in loss of resources and human life and abandonment of territory. As late as the mid-nineteenth century Comanche marauders penetrated as far south as Durango, San Luis Potosí, and some raiding parties reached the state of Jalisco. Little deterred by the Mexican military, the Comanche threat was finally overcome by United States troops and the Texas Rangers.[45]

Some writers have suggested that the position of the present United States–Mexico border in large part may have been determined by the action of nomadic Indian bands that stopped and forced back Spanish settlement in southern Arizona and Texas.[46] Speaking of Arizona, geographer Carl Sauer opined:[47] "The American-Mexican boundary is due more largely to the barrier which the Apache formed from the Gila River to the Texas plains than to any other cause. No Spanish settlements were made in Apache territory, even the frontier garrisons being placed well to the south of the Apachería in the land of sedentary Indians such as the Pima, Opata and Concho."

CHAPTER 7

The Sonoran Gold Craze

Beset by Indian warfare and declining silver production, during the last half of the 1700s many Spanish Sonorans, mixed bloods, and Indians turned to the panning of gold from placer deposits. Some deposits occurred in the sands and gravels along arroyos tributary to the main rivers in La Serrana, the eastern part of the province. The most productive placers, however, were found in the Altar Desert of northwestern Sonora, where the first authentic gold rushes in North America took place during the last quarter of the eighteenth century and continued into the nineteenth century. Except for the placering activity in New Spain during the early sixteenth century, little gold was produced in Mexico until the placer and lode discoveries of the precious metal occurred in Sonora and Sinaloa, beginning in the mid-1700s.[1]

Unlike silver or gold lode deposits, whose exploitation required the investment of substantial capital for equipment and labor and led to the formation of relatively permanent settlements, the working of placer deposits was often a one-person operation or at most a small group association. Since placering required simple tools and nuggets of pure gold needed no further refinement, little or no capital investment was required. Thus, the period of gold placering in Sonora involved many independent operators—common people, including Indians. Moreover, because deposits were small in extent and rapidly worked out, placer camps were ephemeral in nature, at best lasting only a few years. Once the workings were depleted the miners rushed to a newly discovered deposit to repeat the process. Consequently, in Sonora and elsewhere few permanent settlements developed from the exploitation of placer gold.

Gold Placering in Eastern Sonora

The placers of San Antonio de la Huerta were discovered in 1759 on the middle Río Yaqui, and for a time San Antonio was the largest, most-productive, and persistent of the early gold camps in eastern Sonora.[2] Rich deposits occurred in a deep arroyo on the west bank of the Yaqui, upriver from the mission of Tónichi.[3] The discovery drew a rush of poor *gambusinos* (free workers who combed mine tailings) bent on accumulating wealth or at least the wherewithal to purchase food; within a year several thousand Indians and mixed bloods, including some three thousand Yaqui, entered the area to pan the gravels with the simple wooden *batea*.[4] The total amount of gold extracted is unknown, but it must have been large, for some operators were reported to have registered nearly a hundred pounds of nuggets and dust within a few years.[5] According to one report, in 1770, about ten years after the first rush, San Antonio still maintained twenty-five stores, and each year four hundred to five hundred mules loaded with imports from central Mexico and Spain entered the camp to supply the gold seekers.[6] At that time the *alcalde mayor* of Sonora had made San Antonio his headquarters, possibly to monitor the registry of gold and to control smuggling.[7] The bonanza endured longer here than in most placer camps, but with the gradual depletion of its gold-bearing gravels, San Antonio de la Huerta was nearly abandoned thirty years after its founding.[8]

Many other gold placers, all smaller than San Antonio de la Huerta, were exploited in eastern Sonora during the last half of the eighteenth century. Among the more important was Bacoachi (discovered 1758) on the upper Río Sonora, where the presidio manned by Opata Indians was established, probably more to protect the gold workers (mainly Yaqui) than to stem the Apache tide.[9] Farther south within the district of the old Real de San Miguel Arcángel, the placers of Ventanas, which covered an area of four square leagues, were being exploited by Indians later in the century.[10] Again, Saracachi, a mining district containing both gold placers and lode deposits of silver and gold, located in the upper Río San Miguel drainage, was opened for exploitation in 1757; many observers judged it to be one of the most important producers of precious metals in Sonora at that time, on a par with San Antonio de la Huerta.[11] In 1768, 137 families of Spaniards and mixed bloods who inhabited Saracachi were being served by nine stores;[12] the persistence of the settlement may have been due to the working of both placer and lode deposits, and its compactness probably discouraged attacks by Apache and Seri marauders. However, two years later falling output from the mines caused most of the inhabitants to seek new fortunes in a newly discovered gold placer area called Alamillo, twelve leagues away.[13]

By 1770 Saracachi was practically abandoned, permitting the Apache to move in to destroy what was left of a once-prosperous mining center. In 1772 Fray Antonio de los Reyes, the Franciscan bishop of Sonora, opined that the ephemeral character of the Sonoran mining centers of the day was due not as much to falling production from mineral deposits as to the ambition and greed of the inhabitants, who abandoned one mining district for another largely on the basis of rumors of great riches to be had in a newly discovered placer area.[14]

In eastern Sonora gold nuggets and dust were extracted from the sands and gravels along stream banks by means of primitive, time-honored techniques. The miners used wooden hoes and dibbles (*almocafres*) for digging along stream banks. The shallow *batea*, usually made from the lower part of a gourd, served to wash out the worthless detritus; the heavy gold particles mixed with iron oxide grains were left at the bottom of the receptacle.[15] Because of its high price and scarcity, mercury was rarely used to further refine the gold.[16] In places, some two or three meters below a terrace surface, gold-bearing sediment was also found occasionally within ancient buried streambeds, but the material had to be hauled to the nearest stream for washing.[17]

As late as the 1790s belief in the old adage claiming that alluvial gold was always found near the surface because of the influence of the sun's rays was widespread among the local populace.[18] Consequently, few attempts were made to excavate deeply to find the precious metal. Taking advantage of the light cool season rains, or *equipatas*, and low water in the streams the poor people worked placers mainly in winter. High water during summer discouraged placering, and at that time most gold seekers were engaged in agriculture.[19]

During the eighteenth century in New Spain, gold was sold or traded at the rate of ten silver pesos per ounce,[20] hence, the attraction of gold placering over the drudgery and expense of mining and refining silver ores, especially for the lower-class families, which preferred to work individually or in company with others. According to a report of 1762,[21] by law all gold produced was to be registered with local officials, who admitted that much of the metal was traded directly to merchants or mule drivers and smuggled out. In that year Sonoran officials reported that 607 ounces of gold

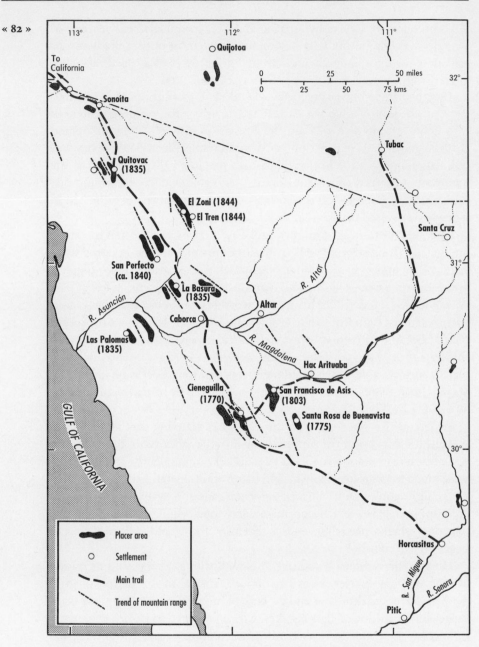

Gold placer areas, Sonoran Desert, eighteenth and nineteenth centuries.

FIG. 22

had been registered, as compared with only 648 ounces of silver, indicating the rising importance of placer mining in the province. At that time gold and silver were carried by mule train, to either San Felipe de Chihuahua (present-day Chihuahua City) or Alamos for assaying.

Gold Placering in Western Sonora

In November 1770, a group of mule drivers accompanying a contingent of Spanish cavalry in pursuit of Seri marauders in the southern part of the Altar Desert discovered a number of large gold nuggets on the sandy surface near their campsite, at a place called Cieneguilla (fig. 22).[22] Shallow excavations revealed even more gold beneath the surface. Once news of this accidental discovery spread to other parts of the province, hundreds, and eventually thousands, of gold seekers flocked into the area in a veritable rush, leading to the first series of settlements and the beginning of Spanish civil authority in that part of arid western Sonora. In the years following, many other rich alluvial gold deposits were found in the Altar Desert; as one deposit played out, the miners rushed to exploit a new find.

This kind of activity continued into the nineteenth century and, in places, well into the twentieth as far north as southwestern Arizona and southeastern California. Although characterized mainly by the formation of ephemeral camps, these gold rushes represented the initial development of western Sonora, which today, not through mining but through irrigated agriculture, has become the leading economic and demographic area of the state.

Geology and geomorphic forces made this particular part of Sonora one of the major storehouses of alluvial gold in Mexico. As described earlier, the surface of the Altar Desert is characterized by remnants of ancient pre-Cambrian mountain ranges eroded by wind and rain to their very bases. From the base of the mountain remnants (inselbergs), the weathered detritus spreads outward as alluvial fans and pediments, forming wide, sloping plains, or *bajadas*. Gold in the form of nuggets and flakes, eroded from quartz veins in the former mountains, has been deposited in the *bajadas*, either on or near the surface or buried more deeply within ancient streambeds now covered with layers of consolidated gravel, or caliche. Both types of alluvial deposits were exploited during the late eighteenth century and beyond by means of simple techniques devised by Indian and mixed-blood workers.[23]

To exploit the surface deposits holes were dug to depths of a few cen-

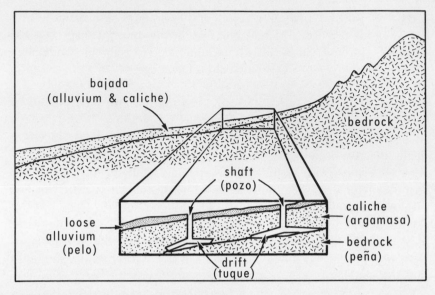

Geology and mining techniques for dry placering.

FIG. 23

timeters to one or two meters. To reach gold-bearing material at depth the workers perfected a simple but effective method of extraction. On most *bajada* slopes a caliche layer underlies the surface sands and consists of gravels, possibly as old as Pleistocene, loosely cemented by lime solutions through time. At the bottom of such layers, three to ten meters below the surface near bedrock, gold occurred abundantly in places, and by digging vertical shafts from the surface through the caliche to bedrock and then by drifting horizontally, the workers were able to recover gold-rich material. In a report written in 1772, a Spanish official on duty at one of the placer camps described this technique, called "de labor" (fig. 23).[24] He also indicated that the caliche deposits were much more productive than the surface gravels. This same technique was used widely in the Altar Desert as late as the first part of this century[25] and still is employed there in a few places. Where the shaft penetrates several meters below the surface, windlasses are employed to haul up detritus for further separation of gold and gravel.

Because little water was available to wash the detritus, colonial gold seekers in the desert developed a method called "dry placering." At first it consisted simply of placing sand and gravel in a wooden *batea* and repeatedly tossing the material in the air to permit wind to remove the lighter substances and leave the heavy gold and black sand (iron oxide) at the bottom of the *batea*, much as one might winnow wheat.[26] Using this primitive method most of the finer gold dust was lost, but nuggets (*oro a granel*) were recovered; the larger ones (some weighing several ounces) were picked out previously by hand. One Spanish official thought that at least half of the gold in the deposits was composed of small flakes, or dust, all of which was wasted by using the winnowing technique.[27]

In March 1771, four months after the initial gold discovery at Ciene-guilla, about fifteen hundred gold seekers and merchants swarmed into an area six leagues (twenty-four kilometers) in circumference, some twelve leagues (fifty kilometers) south of the presidio of Altar (fig. 22).[28] Most of the early migrants (Spanish families, mixed bloods, and Indians) came from the San Miguel and Magdalena river valleys to the east and north.[29] A year later Spanish officials estimated that the population of the small area had risen to five thousand or more, including large numbers of Yaqui Indians, perhaps as many as three thousand.[30]

The number of Yaquis dry panning gold in the Altar Desert was ex-traordinary, even considering the propensity of the males to seek a liveli-hood outside their delta homeland. According to some reports, in Septem-ber 1770 copious rains in Sonora and Sinaloa (a Pacific hurricane?) wiped out harvests, and from 1771 through 1772 flooding and overflow of the lower Yaqui and Mayo rivers failed to materialize, severly limiting the maize harvest in the delta pueblos, where the farmers depended on natural flood irrigation.[31]

Consequently, during those years of flood and famine the Yaqui and probably many Mayo Indians were forced to seek work elsewhere. Once involved in dry placering in the Altar, the Yaqui continued in that occu-pation for generations, well into the nineteenth century. They became adept at prospecting for new placer deposits and may have been a key element in the continuation of gold production in western Sonora for nearly 150 years. One Spanish official summed up the value of Indian work-ers in the Altar Desert thusly: "If Indians of all nations who congregate here should leave, all work would cease; it is they who seek out [gold] deposits with good success. . . . The Spaniards here are useless in such work, for they habitually seek to rest in the shade."[32]

In part to maintain public order, to guard against Seri and Apache raids, and to monitor the production of gold, a formal pueblo, called San Ildefonso de Cieneguilla, was located near a small spring containing brack-ish water in the center of this gold field.[33] Water for human and animal consumption was critical for the continuance of placering, and except for a few wells dug near the spring, water was often hauled in barrels from the Río Magdalena twelve leagues distant at the then-exorbitant price of one peso per mule load.[34]

Life was not pleasant in the Altar Desert. Not only water, but virtually all food was brought in from the Magdalena, San Miguel, and Sonora river valleys and sold at high prices. During the summer months, from the end of May through September, most mining activity ceased, mainly because of the oppressive heat. (Today daytime summer temperatures of 45° C, or 120° F, are not uncommon in parts of the desert.) Moreover, as mentioned earlier, in all parts of Sonora mining stopped during the summer period to allow Indian workers to return to their pueblos to plant and harvest crops and to attend village festivities.[35]

Despite the eight-month work year, the transient nature of the deposits, and the relatively crude placering techniques, the production of gold in the Altar Desert was indeed phenomenal. Within four months after the initial discovery at Cieneguilla, one thousand *marcos*, or eight thousand ounces, of gold gleaned from the surficial sands and gravels had been registered with local Spanish officials; that amount did not include the gold smuggled out to avoid payment of the royal severance tax.[36] At that time an ounce of the precious metal was equivalent in value to ten silver pesos, but at today's price the eight thousand ounces would be worth about three million dollars. The bonanza at Cieneguilla lasted around four years, each year averaging a production of sixteen thousand ounces, or one-half ton of gold—twice the output of all other gold placers in Sonora during that period.[37]

Little government control was apparently exercised over mining claims or labor systems in the Altar gold fields. Labor seems to have been entirely free; because of the lack of permanent native villages close at hand, there was no opportunity to enforce the *repartimiento* system. Indians, mixed bloods, and whites alike appear to have had the right to grub for gold wherever and however they chose. If any Indian worked for wages under prominent white miners, there is no evidence of such in the available documents of the time.

Not all gold seekers struck it rich; many panned little or nothing. Although most nuggets retrieved were small, Indians occasionally found some weighing up to thirty or more ounces, but most of those proved to be impure.[38] Indians usually traded their daily or weekly take of gold for food and other necessities from traveling merchants (*comerciantes*) and small shopkeepers (*tendejoneros*) who always followed in the wake of the rushes. In 1772, according to a local official census, in San Ildefonso de

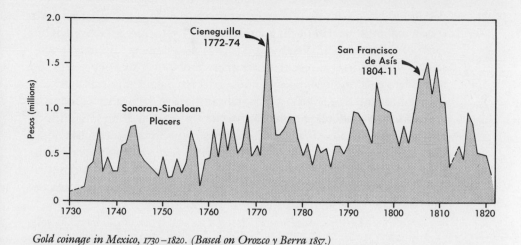

Gold coinage in Mexico, 1730–1820. (Based on Orozco y Berra 1857.)

FIG. 24

Cieneguilla there were twenty-two traveling merchants, nine shopkeepers, and some seventy other whites and mixed bloods—*rescatadores* who bartered low-valued items for gold from the Indians.[39]

The Spanish government provided the local treasury official with silver pesos to purchase gold from merchants and from the more prominent white miners in an attempt to control the export of the precious metal.[40] Once a sizable amount of gold had been accumulated, the metal was boxed and taken by mule train to the *caja real* (royal counting house), at either Alamos or Chihuahua, under military escort, for the levying of the severance tax (usually 10 percent).[41] Before shipment to the mint at Mexico City, gold was usually melted down into *tejos*, discs weighing fifty to seventy ounces.[42]

Rise and Fall of Gold Centers in the Altar Desert

Spanish officials reported a decrease in gold production at Cieneguilla as early as 1773, and two years later expressed fear of its abandonment. At that time, however, a new area of dry placering developed eight leagues (thirty kilometers) east of Cieneguilla at a place named Santa Rosa de Buenavista (also called Palo Encebado).[43] Initially worked by eight hundred Yaqui, Santa Rosa soon was worked by two thousand. In 1776, however, Seri and Apache raids, which had been increasing in frequency and intensity, forced a partial abandonment of the new gold field.[44]

« 88 »   Although exploitation of the placers within a wide radius of Cieneguilla continued until the end of the century, gold production gradually fell because of depletion of deposits and Apache raids on incoming mule trains; serious food shortages resulted and many Indian workers departed. The marauders concentrated most of their atrocities along the trail between Cieneguilla and the Río Magdalena, where the large hacienda of Arituaba, upstream from Caborca, furnished much of the grain and animals destined for the gold fields.[45] In 1791 a Spanish official described Cieneguilla and its surroundings as being "en suma decadencia," with only about one hundred *gambusinos*, despite operation of several gold and silver mines in the mountains nearby.[46] Ten years later more silver mines than gold placers were being worked in the vicinity by wage laborers.[47]

In October of 1803, a new gold field was accidentally discovered only six leagues (twenty kilometers) northeast of Cieneguilla.[48] Named San Francisco de Asís del Oro, this area yielded twenty-six hundred ounces of dry-placered gold within three months of its discovery, and a year later contained a population of forty-five hundred, mainly Yaqui Indians, who with other *gambusinos* maintained the bonanza for several years, with annual yields equal to those of Cieneguilla during its apogee (Appendix E).[49]

Figure 24 shows the role of the Altar Desert placers in the amount of gold coinage in Mexico during the latter part of the colonial period. Concurrent with the San Francisco de Asís bonanza, placers around Cieneguilla again came into production mainly through the efforts of the Indian workers, who began to rework old deposits and discover new ones, thereby establishing a pattern of exploitation that would recur sporadically for the remainder of the nineteenth century and into the twentieth. In early 1806, the combined work force of the two gold fields had grown to more than five thousand; the two camps boasted forty-two small food stores and *cocinas* (field kitchens) and ninety-seven traders and merchants.[50]

Gold production and population gradually fell in the Cieneguilla–San Francisco de Asís fields after 1812 because of decreasing yields and increasing Apache raids on supply trains. The Mexican wars of independence (1810–1824) apparently had less of a negative effect on mining in western Sonora than in the eastern part of the province. The Cieneguilla placers were still producing in 1835 and 1841, when new gold fields were found in the northern Altar Desert; in both years the Cieneguilla area was abandoned temporarily. Excited by rumors of quick riches, the Yaqui Indians, as well as gold seekers from other parts of Sonora, rushed northward to exploit the newly discovered placers in the *bajadas* and gold-bearing

Present-day apparatus for dry placering near Quitovac, Altar Desert. The windlass is used to lift buckets of gold-bearing gravel from a vertical shaft (pozo); the material is sifted to remove the larger cobbles; the finer gravel is then placed in the "Hungarian" dry placer machine (center) to filter out the small gold nuggets and flakes.

FIG. 25

lodes in adjacent hills and mountains. Those of Quitovac, some thirty-five kilometers south of the present Mexico–United States border, were the first to be exploited. The working of other areas followed: La Basura and Las Palomas in 1835, San Perfecto ca. 1840, El Tren and El Zoñi in 1844 (fig. 22).[51] Between 1835 and 1848 gold output from both placer and lode deposits in the northern Altar was estimated to be between two hundred thousand dollars and four hundred thousand dollars U.S. annually.[52]

As production from those deposits decreased, Mexican *gambusinos* pushed farther north into the desert of southwestern Arizona and southeastern California to work small dry and wet placers with the same methods used for generations in Sonora. To be sure, the Quijotoa dry placers in southwestern Arizona had been worked by Pápago Indians since the late eighteenth century, but by the 1850s Yaqui Indians and Sonoran mixed bloods had found several placers along the Colorado River, including the

Pocked surface reflecting years of dry placering near Cieneguilla, Altar Desert.

FIG. 26

Picacho, Pothole, and La Paz deposits north of Yuma and others farther north to the Gila River. These were exploited by both dry and wet methods and eventually were taken over by Anglo miners who worked them until well into the present century.[53]

Sonoran miners were destined to venture even farther north, drawn by still another Ophir. After news of the discovery of gold in California reached Mexico in 1848, some five thousand fortune hunters from Sonora migrated overland and by sea to the new gold fields. These migrants included whites of Spanish descent, Indians (mainly Yaqui) and mixed bloods. They brought with them their modes of dry placering as well as the *arrastre* for milling gold ore. In California, Sonorans spread over the Stanislaus and Tuolumene river basins, and although most had returned to Mexico by the mid-1850s, they left their mark, as reflected by the town named Sonora in the southern part of the Mother Lode Country.[54]

In the 1850s a new mechanical device was introduced into the Altar gold fields to replace the wooden *batea* in dry placering. Called the "Hungarian" dry washer, the machine was a crude air jig that combined bellows and an inclined, cleated oscillating table operated with a hand crank or lever to sift sand and gravel to recover the gold; but, as with the simple *batea*, a large portion of the finer gold particles was lost.[55] Being less labor-intensive than the *batea* to operate, this type of machine is utilized still by the few *gambusinos* who remain in the Altar (fig. 25).

In the late nineteenth century a group of Sonoran mine promoters attempted to remedy the wasteful method of dry placering in the Altar Desert by proposing the construction of reservoirs in a series of arroyos in the arid Los Llanos area near the old San Francisco de Asís fields. They thought that, by damming the sporadic runoff from summer thunderstorms, sufficient water could be conserved to wash the gold-bearing sands so that both nuggets and gold dust could be extracted. Moreover, they also planned to tap the water of the Magdalena River, some twenty-five kilometers to the north, by means of a pipeline and the use of steam-operated pumps. In 1882 the state authorities actually granted the group a concession for the venture, but the estimated several hundred thousand pesos needed for the project could not be raised and the plan was abandoned.[56]

From 1884 to 1894 the Cieneguilla area (today La Ciénega) saw a new bonanza based on dry placering, aided by capital from United States mining companies and the use of the dry washer. As many as ten thousand Yaqui workers once again began to placer mine old deposits and discover new ones.[57] The decade of the 1930s experienced the last rush into the declining Altar gold fields, at present worked by only a few poverty-stricken *gambusinos*, despite the current inflated price of the precious metal.

Today, except for occasional isolated hamlets and ranchsteads, most of the Altar Desert is as desolate and uninviting as it might have been before the onset of the gold period over two hundred years ago. But for several kilometers around old Cieneguilla and other deserted gold camps, the surface gravels are pocked with shallow depressions surrounded by heaps of detritus half overgrown by desert shrubs—a vivid legacy of the frenzied search for quick wealth (fig. 26). Only the irrigated district along the lower Magdalena-Asunción River plus a few fishing ports and recreation areas along the coast have yielded to modern development.

CHAPTER 8

The Growing Domination of Western Sonora

For much of its time under Spanish and Mexican control, Sonora was dominated culturally and economically by its eastern, mountainous portion. During the last half of the present century, however, its arid western part has experienced rapid economic growth, especially in irrigated agriculture within the flattish river deltas. This development has made western Sonora the leading section of the state in terms of population, economy, politics, and associated phenomena. This shift of human endeavor has left the long-settled east as a cultural backwater, characterized by quaint colonial villages in narrow valleys with limited farming potential and, in the adjacent mountain ranges, by remnants of old mining centers that once gave wealth and prestige to the area. Thus, the present-day state of Sonora reveals a dual geographical personality—eastern, colonial, "Old" Sonora, as opposed to western, modern, "New" Sonora (fig. 27).

After only fifty years of relative prosperity in missionizing and mining (1640–1690), eastern Sonora entered a long period of gradual economic and demographic decline that began with the Apache and Seri incursions during the late seventeenth and early eighteenth centuries (see chapter 5). The raids that continued almost unabated for nearly 150 years, coupled with declining mineral production, left much of the province prostrate by the mid-1800s. Although the economy was temporarily revived by the gold rushes of the late eighteenth and early nineteenth centuries, that activity was transitory and failed to lead to permanent settlement in the desert of western Sonora.

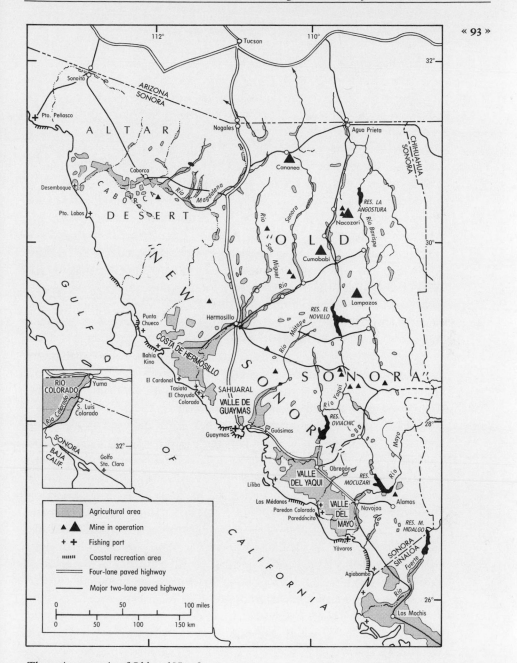

The main economies of Old and New Sonora, 1990.

FIG. 27

Development of Sonora's Main Commercial Axis

The last half of the nineteenth century saw the beginning of an economic and demographic rise within the western part of the state. This growth first involved two small population centers within the eastern margin of the Sonoran Desert: the town of Guaymas, the main port of the state, and Hermosillo, which in 1879 was made Sonora's political capital. The early development of the two centers was based on foreign and local commerce—overseas through the port of Guaymas, by land from Guaymas to Tucson in the Territory of Arizona, United States. By 1860 a wagon road extended northward from the port to Tucson, via Hermosillo, a distance of some five hundred kilometers. Wagon trains hauled foodstuffs, mainly wheat and flour, from Hermosillo into southern Arizona, where North American mining interests and settlers offered a good market. Moreover, wheat and flour were exported through Guaymas to San Francisco and the gold fields of California, as well as to various points along the west coast of Mexico; hides went to England in exchange for mining equipment and other manufactured products.[1] Machinery for Sonoran and Sinaloan mines was also shipped in from San Francisco.[2]

By 1882, following the route of the wagon road, the Sonoran Railroad was completed between Guaymas and the United States border. A customs station and the town of Nogales were established on the border, thereby modernizing the north-south commercial axis of Sonora.[3] Acquired by the American-owned Southern Pacific Railroad Company in 1905, the line was eventually extended southward to tap the growing agricultural areas in the Yaqui and Mayo deltas in southernmost Sonora, the Fuerte delta of Sinaloa, and the Pacific coastal lowlands as far as southern Nayarit; in 1924 the line finally joined the Mexican National Railroad line to Guadalajara on the plateau of central Mexico.[4] Today the infrastructure of the commercial axis of Sonora consists not only of the railroad, but also of a paved four-lane divided highway from Tucson, Arizona, to Navojoa in the Mayo delta and beyond. For most of this century this axis of communication and trade has been a major factor in the development of much of northwestern Mexico as well as the western part of Sonora.

Well protected from storms by surrounding mountains and offshore islands, the extensive bay of Guaymas is the best natural harbor in northwestern Mexico. As early as the eighteenth century, supplies for the Baja California missions occasionally were shipped from the bay, and after independence from Spain the small port was opened to foreign commerce.[5] During the nineteenth century its role as Sonora's main port was well established by the presence of several large Mexican, North American, and English mercantile firms, a regularly scheduled North American steamship line to San Francisco, and as a port of call for other foreign vessels.[6] By 1864 the population of Guaymas and vicinity numbered six thousand, and at that time it was one of the larger towns of the state.[7]

In 1700 the site of Hermosillo was occupied by a small settlement called Pitic, housing a few Spaniards and Pima Bajo Indians and located along the Río Sonora at the edge of the western desert.[8] For a time it served as a presidio against the rebellious Seri Indians, but eventually the site became the center of an hacienda. In 1780 the *villa*, or town, of Pitic was formed nearby and eighteen Spanish *vecinos* received small plots of land and water rights for farming.[9] Thereafter the irrigated fields along the lower Río Sonora near the confluence with the Río San Miguel, upstream from the *villa*, became a flourishing wheat-producing area. It was renamed Hermosillo (in honor of a Mexican hero of the War of Independence) in 1828, and by the mid-nineteenth century the *villa* had become the largest and most prosperous town in Sonora with a population of around fifteen thousand.[10] By that time it had developed its role as a collection point for commodities (wheat and hides) from the river valleys of La Serrana; these were exported through Guaymas or hauled northward to Arizona. Hermosillo also became a distribution center for foreign goods carried in from the port. Some twenty-five mercantile establishments in the town handled the trade, and several flour mills were constructed as some of the first industries of the area to process agricultural products on a large scale.[11]

After the Gadsden Purchase in 1853, Tucson, the northern terminus of the Sonoran commercial axis, became a part of the United States. Nogales, on the new international border, was not established as a town until 1884, two years after the completion of the Sonoran Railroad.[12]

Revival of Mining, 1880–1910

After the final subjugation of the nomadic Indian menace in the 1880s and the establishment of political stability in Mexico with the inauguration of Porfirio Díaz's dictatorship in the late 1870s, the fortunes of eastern Sonora temporarily improved. During the Porfiriato (1877–1911) North American companies began to invest heavily in Mexico's mining industry. In the mountains of La Serrana foreigners reopened many of the abandoned silver, gold, and copper mines, called *antiguas*, and used modern techniques to exploit the deep, poorer ores. According to one report, by 1884, eighty mining companies, the majority North American—owned or financed, had reopened some two hundred old colonial gold and silver mines, mainly in eastern Sonora.[13] Included were those in the vicinity of seventeenth-century Nacatóbari (Lampazos) east of the Moctezuma River; many small deposits on either side of the Río Sonora Valley; and those of San Javier near colonial San Miguel Arcángel. And in the Alamos district a British concern began to rework the famous La Quintera mines near La Aduana and Minas Nuevas.[14] Such operations involved the use of the traditional amalgamation process as well as the new leaching methods (for example, cyanidation) of refining ores, plus the employment of steam-powered stamp mills,[15] thereby reviving the moribund mining industry of eastern Sonora, if only for a relatively short time.

During this period, however, the most important mining development in the state entailed the exploitation of low-grade copper ores at Cananea and Nacozari in deposits that form the southern end of the great copper belt of Arizona. Organized in 1899 by financier and rancher William C. Greene, the Cananea Consolidated Copper Company, together with other North American–owned mining concerns, by 1906 employed 2,250 workers.[16] At that time the Cananea area, with 20,000 inhabitants, formed the largest population center in the state.[17] By 1910, at Nacozari, the old colonial silver mining area, the Moctezuma Mining Company was producing substantial amounts of copper, gold, and silver.[18] Both Cananea and Nacozari remain Mexico's principal copper mining centers.

The revival of Sonoran mining was short-lived; it was suddenly arrested by the outbreak of the Mexican Revolution in 1910. Excepting the copper mines at Cananea and Nacozari and the recent exploitation of a few barite, graphite, zinc, and molybdenum deposits for industrial use, the mining industry has never wholly recovered in eastern Sonora. During the early twentieth century western Sonora continued to develop in commerce and especially in irrigated agriculture, which has led to its present dominant position in the state and its particular geographic personality, so different from that of its eastern counterpart.

Early Mexican and North American Involvement in Irrigated Agriculture in Western Sonora

Beginning slowly in the 1890s, the development of irrigated commercial farming in the river deltas of Sonora and Sinaloa was by far the most significant factor in the modernization of that part of northwestern Mexico. The completion of the Sonoran Railroad in 1882 and the possibility of its extension southward from Guaymas occasioned the hope among many Mexican and North American promoters that Sonora and Sinaloa might one day become important exporters of agricultural products to the United States; moreover, a final extension of the railway along the Pacific coastal plain to Guadalajara on the plateau would tap the large market of populous central Mexico. Such optimism led to plans to develop large-scale irrigation within the extensive delta plains along the coast—first, the Yaqui and Mayo river deltas in southern Sonora, as well as that of the Río Fuerte and others farther south in Sinaloa; later proposals were made to develop the delta of the Colorado River in northern Baja California; and finally, in the mid-twentieth century, irrigation of the deltas of the Río Sonora in the west-central part of the state, and farther north, that of the Río Asunción–Magdalena both came under governmental planning. The early developmental schemes, however, especially those made in the late 1800s and early 1900s, often led more to land speculation than to true agricultural growth.

The fertility of the delta soils was well known. Even the famous geographer Alexander von Humboldt considered the lower river plains of northwestern Mexico to be among the most potentially productive in all of New Spain.[19] But the perception of such fertility was based largely on the agricultural production of the sandy loam soils within the floodplain immediately along the rivers that annually overflowed, with the deposition of rich organic sediment as natural fertilizer. However, in the mesquite- and cactus-covered part of the delta plain, away from the active rivers, the heavy clay soils were less fertile, often difficult to cultivate, and in places moderately saline—detrimental features that were discovered once the land was cleared and prepared for irrigated farming.[20]

Another problem stemmed from the seasonality of river flow. Along the larger rivers control of the annual summer floods that often destroyed diversion dams, water intakes, and canals was initially beyond the ken and wherewithal of the early land developers. Moreover, decrease in river flow during the long dry season prevented year-round irrigation. Only after such problems were overcome by constructing large reservoirs during the 1940s and 1950s to control the distribution of river water was the full agricultural potential of the delta plains finally realized. Water and its control was the key to the modern development of arid western Sonora.

The early attempts to develop the deltas of the Yaqui and Mayo rivers were delayed by rebellious Yaqui and Mayo Indians, who feared that foreign and Mexican settlers would occupy their traditional farmlands. Especially the Yaqui fought both government and private plans to open un-

« 98 » occupied portions of the delta to settlement. The Indians considered as their inviolable homeland not only their farms along the Yaqui River, but also the entire delta, including the extensive plain south of the river as well as the hills and mountains (especially the Bacatete Range) to the north, their traditional area of refuge. Sporadically throughout much of the last half of the nineteenth century and into the first decade of the twentieth, bands of rebellious Yaqui fought the Mexican military and ravaged Mexican and foreign settlements within and on the borders of what they claimed as their territory. So unremitting was Yaqui resistance to the government that some officials declared a campaign of extermination; between 1902 and 1904 hundreds of men, women, and children were moved to the henequen plantations of northern Yucatan and the sugar haciendas of eastern Oaxaca, where they were treated as virtual slaves. To escape imprisonment many Yaqui fled to the United States. Toward the end of the Porfiriato (1910) the Yaqui who remained in Sonora finally made temporary peace with the Mexicans, and development of irrigated farming in the delta continued apace.[21]

As early as 1858 the federal government formulated plans to colonize the Yaqui and Mayo deltas with refugees from Alta California who wished to avoid living under United States rule.[22] Because the Indians threatened rebellion, these plans were delayed, but ten years later some concessions were doled out to Mexican settlers along both rivers with promises to construct canals to lead water to their holdings. Little more was done until the 1880s, when the government organized a Comisión Geográfica composed of army engineers to survey and distribute land in the deltas. Although a few Mexicans were settled along the middle portion of the Mayo above the entrance into its delta, Yaqui hostility prevented colonization elsewhere. Later in the decade other military geographical commissions followed without much progress in settling the deltas, and it was not until the 1890s that more serious attempts at development were made—in this case, by private enterprise rather than through direct government endeavor.[23]

In 1890 a Mexican entrepreneur, Carlos Conant Maldonado, of Alamos and Guaymas, with North American financial backing, obtained from the Mexican government a concession to develop and colonize the deltas of the Yaqui, Mayo, and Fuerte rivers. Canals were to be constructed to distribute river water for irrigating a total of a half million hectares of land, three hundred thousand hectares of which was composed of the arid, bush-covered part of the Yaqui delta south of the river floodplain.

Under this concession Conant and his Mexican and North American associates organized the Sonora and Sinaloa Irrigation Company, chartered in the state of New Jersey and financed largely by New York interests that controlled 75 percent of the company. Conant, as general manager, and his Mexican partners controlled the rest. This joint Mexican-U.S. enterprise marked the beginning of large-scale development, especially in the Yaqui delta. The company planned to sell commercial farms to North American and other foreign settlers, who would produce crops for both export and Mexican consumption.[24]

During the 1890s the company completed only thirty-six kilometers of a canal that diverted water from the Yaqui River by means of a small dam constructed near a place called Hornos in the upper portion of the delta; the canal led along the south bank of the river and ran adjacent to Indian farms, but the water was solely for the use of white settlers. Canals were also begun along the north bank of the Mayo and both banks of the Fuerte in Sinaloa.

To cope with the seasonality of river flow, Conant planned to store the overflow of summer floods in ponds and natural depressions within the floodplain for use during the dry season. But owing to Indian troubles progress was slow, causing completion of the canals to be much delayed. Moreover, despite the widespread distribution of propaganda in the United States extolling the virtues of commercial farming in the Yaqui area,[25] by 1901 only eight hundred hectares of land had been sold to prospective colonists, forcing the company into receivership.[26]

During the period the Sonora and Sinaloa Irrigation Company was in business, Conant's engineers surveyed the northern part of the Yaqui delta, immediately south of the river, and established a cadastral system unique in Mexico but in many respects similar to the township and range system common in the United States. The nationality of the engineers is uncertain, but they may well have been North Americans. An east-west baseline was laid out along the southern border of the floodplain, accompanied by a north-south prime meridian that today lies some eight kilometers west of the present limits of Ciudad Obregón.[27] (At present, a paved road called Calle Meridional runs along the line.) Based on the two prime lines, square sections of land, each measuring two by two kilometers, were surveyed to form a grid system of land division. Called blocks (*bloques*), each section contained four hundred hectares, initially the basic unit of ownership and land division within the Yaqui delta. Later, to accommodate small holders, each block was further divided into forty lots of ten hectares each.

« 100 »

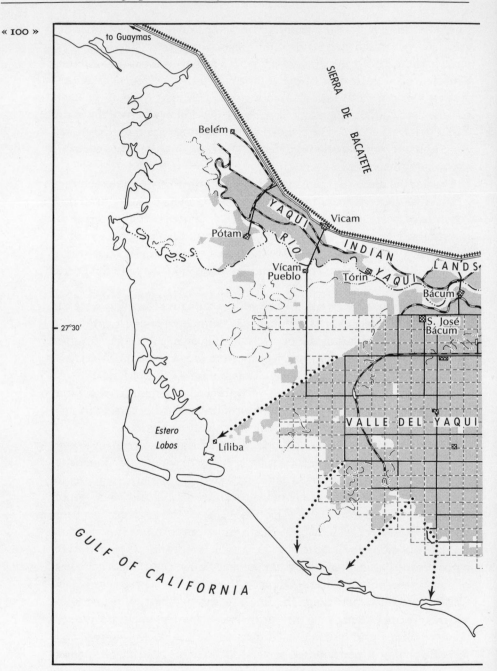

Irrigated agriculture, Valle del Yaqui and northern part of Valle del Mayo, 1980s.

FIG. 28

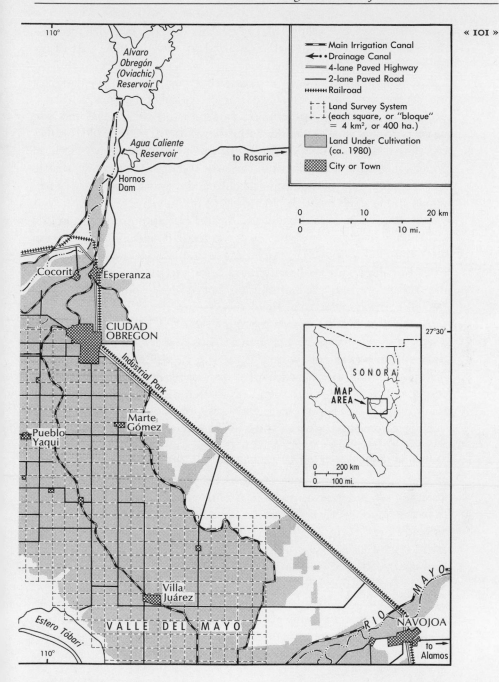

Main Irrigation Canal
Drainage Canal
4-lane Paved Highway
2-lane Paved Road
Railroad
Land Survey System
(each square, or "bloque"
= 4 km², or 400 ha.)
Land Under Cultivation
(ca. 1980)
City or Town

110°

Alvaro Obregón (Oviachic) Reservoir

Agua Caliente Reservoir

to Rosario

Hornos Dam

Cocorit

Esperanza

CIUDAD OBREGON

Industrial Park

Marte Gómez

Pueblo Yaqui

Villa Juárez

VALLE DEL MAYO

Estero Tóbari

NAVOJOA

RIO MAYO

to Alamos

110°

27°30'

SONORA

MAP AREA

0   200 km
0   100 mi.

0   10   20 km
0   10 mi.

« 102 »   This cadastral system still prevails (fig. 28) and has been extended throughout the delta (in the Valle del Yaqui) and southward into a part of the
Mayo delta (the Valle del Mayo).[28]

The square blocks of 400 hectares each appear to be an adaptation of
the "section," one mile square, or 640 acres, typical of the American township and range system; furthermore, the section may be divided into half,
quarter, or even smaller holdings. Apparently, in Mexico the block cadastral system is found only in the Yaqui and part of the Mayo deltas of
Sonora and differs markedly from the large rectangular and oddly shaped
landholdings of the old Mexican hacienda and the smaller, irregularly patterned ejido parcels found in other parts of the country.

With the failure of the Sonora and Sinaloa Irrigation Company, in 1903
Conant and his friends formed a new enterprise (Compañía de Irrigación
del Yaqui) and continued to extend canals and land clearing south of the
river. But the disastrous flood of 1905 destroyed practically all improve-
ments from previous years, causing Conant to retire from agriculture and
land speculation in the delta. He died two years later.[29]

Another attempt to develop the Yaqui delta was made by three North
Americans—the Richardson brothers of Los Angeles—who had become
successful miners in Sonora during the late 1880s. Enticed by the possibilities of adding to their wealth by speculating in land along the west
coast of Mexico, in 1904 the brothers formed the Richardson Construction Company and applied to the Mexican government for rights to take
over developments begun by the Conant organization, but only in the
Yaqui delta. Five years later, armed with fifteen million dollars invested by
American financiers, the Richardsons finally received a concession to pro-
ceed with plans for large-scale irrigation that involved the improvement of
water intakes and the construction of a main canal (Canal Principal Bajo)
from the Yaqui southward into the desert plain.[30] The extension of the
railroad from Guaymas to the Yaqui River (1906) and beyond to the Mayo
afforded a means to export crops and thus greatly increased prospects for
land sales.[31] The new company continued the cadastral survey initiated by
Conant, constructing lateral secondary and tertiary canals along the borders of the four hundred–hectare blocks, thus forming a regular grid pattern of water distribution by gravity flow on the delta plain, which slopes
gently seaward—a system still used today.[32] Ease of transport was assured
by building a network of farm roads that followed the borders of the four
hundred–hectare blocks.

The Richardsons began a vigorous propaganda campaign in California
magazines and newspapers and elsewhere in the United States to sell land

to prospective American settlers.[33] Land was offered in four hundred–hectare blocks or in lots as small as ten hectares, or twenty-five acres,
at twenty-five dollars U.S. per acre.[34] Many Americans took the bait as did
several Europeans, mainly Germans, who emigrated just before or during
World War I. Both American and German settlers in the Yaqui delta introduced modern agricultural techniques—the use of large farming machinery, such as multifurrow moldboard plows, land-leveling machines, the
McCormick binder, and later the combine for harvesting wheat and other
crops—all drawn by tractors. A German, one Hermann Bruss, who settled
in the Yaqui delta in 1910, is said to have been the first in the area to use
an internal-combustion engine tractor to cultivate his land and a combine
to harvest his wheat crop.[35] Much later, foreigners would introduce the
use of commercial fertilizer, insecticides, herbicides, and improved crop-
processing plants. It might be said that through the Richardson concession modern farming techniques—today the hallmark of most commercial
farms in Mexico—were first applied in the country on Yaqui delta irrigated holdings.

Like the promoters who preceded them, the Richardsons were plagued
by various problems, the most serious of which was the irregular discharge
of the Yaqui River; during the dry season and dry years the river furnished
insufficient water for year-round irrigation, and the summer floods always
threatened destruction. Company officers were among the first to propose
construction of large dams and reservoirs upstream from the delta to help
regulate river flow. A canyon called La Angostura on the lower Bavispe
River, a main tributary of the Yaqui, was suggested as a place to construct
a reservoir, but the outbreak of the Revolution postponed the project,
which was not revived by the Mexican government until the late 1930s.[36]
Until the construction of such reservoirs the full utilization of Yaqui delta
land would be impossible. Another recurring problem was the continuing
Yaqui Indian resentment against settlement of their land by Mexicans and
foreigners south of the river. In 1914 rebellion broke out anew, causing the
temporary abandonment of half of the area then under cultivation.[37]

Nonetheless, by 1925 the Richardsons had opened up more than twenty-
five thousand hectares of delta land to irrigation.[38] Except for Yaqui village
plots along the river, nearly all land under cultivation was privately held.
Twenty-five holdings exceeded four hundred hectares, and one individual
owned thirteen blocks, or a total of fifty-two hundred hectares (thirteen
thousand acres), forming a sizable "hacienda."[39] Most of the properties,
however, ranged from ten to four hundred hectares. Within the settled
area (the northeastern part of the delta plain) there were few towns,

« 104 » Pueblo Yaqui with a few hundred inhabitants being the largest; most settlements were small hamlets (*congregaciones*), composed of hired laborers, and scattered homesteads of landowners.[40] In 1910 the railroad station, called Cajeme after a renowned Yaqui leader and located near the northeastern limit of the delta, was a collection of warehouses and company offices. Twenty-five years later this modest hamlet had grown into a town, now known as Ciudad Obregón, of several thousand inhabitants. Today, with some 250,000 people, it is the commercial, industrial, and social center of the Yaqui delta and second only to the state capital, Hermosillo, as Sonora's main urban center.

During the Richardson period (1909–1926) the main crops produced by irrigation in the Yaqui delta were wheat, rice, and maize. As in most of Mexico, wheat was the winter crop, because the rust diseases rampant during the hot, showery summers could be avoided. Approximately half of the crop was milled and consumed locally or shipped to other parts of the country; the other half was exported via Guaymas or other Pacific ports. Rice and maize were the principal summer crops, the former thriving under irrigation on the delta's heavy clay soils, which retained water near the surface. Maize, the main crop throughout Mexico, was used locally, but most of the rice was exported through Guaymas to Canada and Europe. Minor crops included melons, shipped by rail mainly to California, beans, lentils, some cotton, and a few vegetables such as lettuce and green peppers sent to markets in the American Southwest.[41] Thus, by the 1920s the export function of irrigated agriculture along the west coast was fairly well established.

The fact that most of the irrigated land in the Yaqui delta was farmed by Americans and Europeans (a practice said to have been encouraged by the Richardson brothers) may have caused the Mexican government to suspect that the company purposely discouraged Mexican farmers from purchasing land and settling as colonists. In 1926, in the aftermath of the Revolution, when land reform was gradually sweeping the country, the government forced the Richardsons to sell controlling interest of their company to the newly formed Comisión Nacional de Irrigación,[42] an action that began the conversion of agricultural development along the Mexican west coast from private to government control.

In contrast to the development of irrigated commercial farming in the Yaqui delta, during the early twentieth century the scrub-covered portion of the Mayo delta to the south was in large haciendas, owned and operated by Mexicans raising horses and mules. Along the river floodplain Mayo Indians and Mexican small farmers were densely settled in ten pueblos and more than one hundred hamlets; they irrigated fields year round with water stored in *bolsas*, or natural depressions, filled from river overflow during summer floods. Maize, wheat, and other crops were cultivated mainly for subsistence rather than for sale.[43] Never as bellicose as their Yaqui neighbors, Mayo Indians for the most part lived peacefully with Mexicans and miscegenation was not uncommon. Large-scale irrigation in the Mayo delta was not undertaken until later, when the Mexican government began operation there under the auspices of the Comisión Nacional de Irrigación.

In the neighboring state of Sinaloa, however, by the early 1920s commercial farming was well under way in the Fuerte River delta, where irrigated and rain-fed agriculture was financed and developed largely by North American capital from California and Arizona. For example, in 1925 the American Sugar Company had forty thousand hectares not only in sugar-cane, but also in tomatoes as a winter crop exported to the United States.[44]

The Role of the Mexican Government in Irrigated Agriculture in Western Sonora

During the Porfiriato the federal government left the development of irrigation in Mexico in the hands of national and foreign individuals or companies—a policy that proved to be unsatisfactory in terms of national economic and social goals. Thus, as a result of the 1910 Revolution, in 1926 the administration of Elías Calles promulgated the Water Law (Ley de Irrigación con Aguas Federales), which committed the federal government to developing large-scale irrigation projects throughout the country. In some respects the new law was patterned after the United States Reclamation Act of 1902.

To administer the law the government formed the Comisión Nacional de Irrigación, the main duties of which were to make studies for selecting areas best suited for irrigation; to undertake the construction of large-scale dams, reservoirs, and canals; and to colonize the newly formed irrigated districts. To finance the large costs of construction, the Fondo Nacional de Irrigación was set up to administer federal funds, but owners of irrigated land were obliged to help by constructing and maintaining minor works such as water intakes and lateral canals, and were required to pay a tax on land irrigated with federal water.[45] The goals of the whole program initially consisted not only of increasing agricultural production for both home consumption and export, but also of creating a middle class of small

« 106 » landed farmers who would have the incentive to produce more than the poorer subsistence class.[46]

The first twenty years of the Comisión Nacional de Irrigación were devoted mainly to collecting data and planning the establishment of irrigation works. Because of limited federal funds few large-scale dams and reservoirs were constructed, and those that were completed served to control seasonal river discharge and to lessen flood damage.[47] In 1946 the Comisión was replaced by the Secretaría de Recursos Hidráulicos, which had cabinet status in the six-year presidential administrations and thus, through increased access to federal funds, was able to inaugurate a period of impressive dam and reservoir construction and expansion of irrigated lands in many parts of the country.[48]

Developments in the Yaqui and Mayo Deltas

In Sonora the initial efforts of the Comisión Nacional de Irrigación to improve water supply and expand irrigated farming were concentrated in the Yaqui delta, the largest agricultural area of the state and an area in which private enterprise had already made significant headway. Following plans previously made by the Richardson Company, the Comisión in 1936 began construction of a large dam and reservoir at La Angostura on the Río Bavispe. As a major tributary of the Yaqui, the Bavispe and its affluents drained a large portion of the northern Sierra Madre Occidental, where heavy rains from July through October supplied much of the runoff that caused summer floods in the delta; in the dry season (April through June), however, it furnished little water for irrigation.

Completed in 1942, La Angostura reservoir served to regulate river discharge so that water was available for year-round irrigation. This availability led to construction of new major canals and the doubling of cultivated area in the delta to nearly 115,000 hectares. One of the new canals led along the northern bank of the river and furnished water for the Yaqui Indian village lands, theretofore dependent on annual summer floods for their subsistence crops.[49]

Later, under the supervision of the Secretaría de Recursos Hidráulicos, even more impressive developments occurred in the delta. In 1947 construction began on the Oviachic dam and reservoir (Alvaro Obregón) on the lower Yaqui, where it debouches onto its delta thirty-five kilometers north of Ciudad Obregón. This, the largest reservoir (67,000 square kilometers) on the Yaqui system, on its completion in 1952 made water avail-

*One of the main canals (*acequias madres*) used to distribute water to fields in the Valle del Yaqui. The lock in the foreground controls the flow.*

FIG. 29

able to irrigate 220,000 hectares, half of which, still covered with desert scrub, was quickly cleared with large bulldozers.[50] The block survey system, started more than a half century earlier, was rapidly extended into the newly cleared land, lateral canals were excavated, and crops were planted on farms that reached southward to the edge of the Mayo delta (figs. 28, 29). Finally, in 1963, a third dam, called El Novillo (now Elías Calles) was completed at the confluence of the Yaqui and Moctezuma rivers midway between the Oviachic and Angostura reservoirs. Constructed mainly to generate electrical power for the Yaqui and Mayo irrigation districts, the Novillo reservoir also serves to regulate further the water regime of the Yaqui delta.[51]

Small amounts of irrigation water have also been obtained from deep wells that tap aquifers within the thick deltaic alluvium. In order to supplement river water during the dry season, such wells, all drilled by private land-owners, were especially important before the completion of the

« 108 »   Oviachic reservoir. But because of the limited extent of the aquifers, the wells furnished water only for a short time.[52] Today only about fifty are in use.

The Mayo River delta, smaller in area than the Yaqui delta, did not receive the full attention of the federal government's irrigation agencies until the 1940s. As mentioned earlier, the Conant enterprise had made small diversion works in the late 1890s, utilizing the summer floods to irrigate the upper portion of the delta. The Comisión Nacional de Irrigación began studies on the river basin as early as 1928, and, finally, ten years later, major canals were constructed to bring water into the central part of the delta, making some twenty thousand hectares (later doubled) available for irrigation.[53] Like the Yaqui River, however, nearly 70 percent of the annual discharge of the Mayo occurs from July into October, resulting in frequent summer floods and lack of water for irrigation during the dry season.

To regulate river discharge, thereby making water available for year-round cultivation, and to generate electrical power, in 1952 the federal government began construction of a large dam at Mocuzari, forty kilometers northeast of Navojoa; it was completed in 1955. The reservoir, named after President Adolfo Ruiz Cortinas, made possible the irrigation of an additional thirty thousand hectares of desert land, bringing the total cultivated area of the delta to more than seventy thousand hectares.[54] After clearing the land formerly in haciendas, the new farming area was divided into smaller units by extending the block survey system used in the Yaqui area southward to include the northern part of the Mayo delta. The central and southern sections of the delta, largely in small farms long occupied by Mayo Indians and Mexican mestizos, were not resurveyed and still retain their irregularly shaped properties. Today irrigated farming is continuous from the Yaqui River to well south of the Mayo and joins the two deltas to form the largest single agricultural area along the west coast of Mexico (fig. 27).[55]

During the 1960s signs of soil salinity were appearing in both the Yaqui and the Mayo deltas. Excessive application of water to fields and the rise of the water table resulted in poor drainage and were thought to be the main cause of increasing salinization. The problem was less advanced in the Mayo area than in the Yaqui, where by the late 1960s nearly 13 percent of the irrigated area near the coast was affected. There, the matter was partially solved by excavating deep drainage ditches to conduct excess water to the Gulf (fig. 28).[56]

Developments in the Deltas of the Sonora « 109 »
and the Concepción-Magdalena Rivers

Located within the most arid part of the state, Sonora and the Concepción-Magdalena deltas present irrigation problems that in many respects differ from those encountered in the Yaqui and Mayo areas farther south. Though the seasonal regime of river flow (summer maxima, winter minima) is similar to that of the southern rivers, annual discharge of the northern streams is far less; as indicated earlier, the flow of neither the Sonora nor the Concepción reaches the Gulf of California, except during exceptional summer floods. Thus, the construction of large reservoirs to contain sufficient water for extensive irrigation projects has not been feasible. However, below the arid surface of both deltas voluminous aquifers have formed at depths of fifty to one hundred meters through gradual seepage of water from the lower portions of the main rivers and desert arroyos over a period of thousands of years, probably during and since the Pleistocene. By drilling deep wells and using electrically powered pumps, these large accumulations of groundwater today supply sufficient water to irrigate most of the two deltas as well as adjacent desert basins, thereby adding to the large agricultural production of western Sonora. Unfortunately, water extraction in the Río Sonora delta (called the Costa de Hermosillo irrigation district) since the 1950s has far exceeded natural replenishment, which has led to a rapid fall in the water table and the intervention of government irrigation agencies to control pumping by private landowners. The case of the Río Sonora delta is a prime example of the "mining" of fossil water for short-term financial gain and, on a smaller scale, is similar to the exploitation of the great Ogallala aquifer that lies beneath the Great Plains of the United States.

Irrigated farming along the Río Sonora in the vicinity of present-day Hermosillo had been established on a small scale since the late eighteenth century; by the mid-1800s Mexican farmers had planted a few plots of wheat and maize along the riverbed as far downstream as the Siete Cerros, a group of hills forty kilometers west of town. West of Siete Cerros the extensive delta plain was devoted mainly to stock raising. One of the first cattle haciendas, called San Francisco de la Costa Rica, was established in 1844 by the Encinas family of Hermosillo in the northern part of the delta. There, in 1898, the first deep well lined with metal pipe was drilled to tap an aquifer, in order to supply water for livestock and irrigate fields of maize and wheat.[57] After 1910 most of the arid delta land was purchased

« 110 » by North American interests. One firm, the R. R. Bailey Company, owned two hundred thousand hectares of desert scrub for raising cattle. Despite their awareness of the existence of plentiful groundwater beneath their land, the owners, with no interest in farming, failed to exploit the resource.[58]

Encouraged by government propaganda to increase Mexico's agricultural production, during the 1940s local landholders, who, according to land reform laws, could own up to one hundred hectares of irrigated land, began boring deep wells to depths of 150 to 200 meters with U.S.-made rotary drilling equipment. In 1945, 15 wells were drilled; six years later the number had reached 258. By the late 1950s water from 450 wells was being used to irrigate more than one hundred thousand hectares of privately owned land, planted mainly in wheat and cotton. By 1980, 887 wells were in use.[59] The Costa de Hermosillo had become second only to the Valle del Yaqui as Sonora's foremost agricultural area (fig. 30).

To supplement the water supplied by wells, in 1949 the federal government completed a small reservoir (named Abelardo Rodríguez) on the Río Sonora adjacent to Hermosillo; it hoped to expand farming along the river as far as the Siete Cerros via a forty-kilometer canal.[60] Today the water stored in the reservoir is almost wholly utilized by the growing city for domestic and industrial needs, which forces the farmers in the entire delta to rely exclusively on wells.[61]

Although the delta's water supply was placed under the supervision of the Secretaría de Recursos Hidráulicos in 1951,[62] it was not until the early 1960s that indications of a rapidly dropping water table were evident owing to the increasing overdraft of water pumped from the region's aquifers. It was estimated that each year two and one-half times more water was being extracted than replenished through natural processes.[63] Moreover, since the water of the Río Sonora, which once helped maintain the groundwater level, was now stored in the Rodríguez reservoir, the few remaining sources of replenishment consisted only of intermittent desert arroyos, such as that of Bacoachi in the northern part of the delta.[64]

Thus, fear of intrusion of salt water from the Gulf of California, increasing costs of pumping from greater depths, and the eventual depletion of the aquifers led to strict government control over the number of wells permitted to extract water for irrigation. In 1989 the Secretaría de Agricultura y Recursos Hidráulicos limited the number of wells to 498 and halved the annual amount of water pumped from the normal eight hundred million to four hundred million cubic meters.[65] Such drastic measures

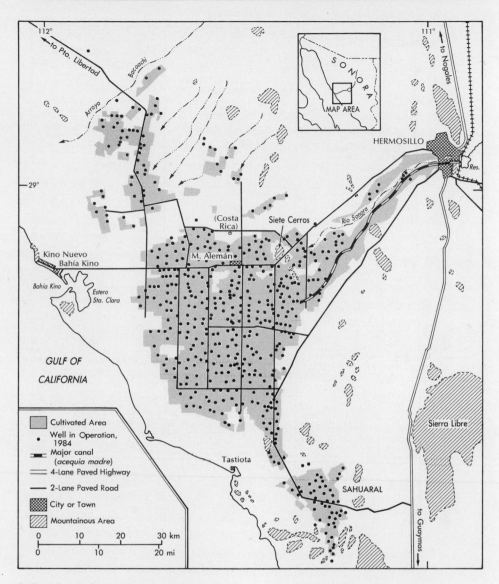

Irrigated agriculture, Costa de Hermosillo, 1980s.

FIG. 30

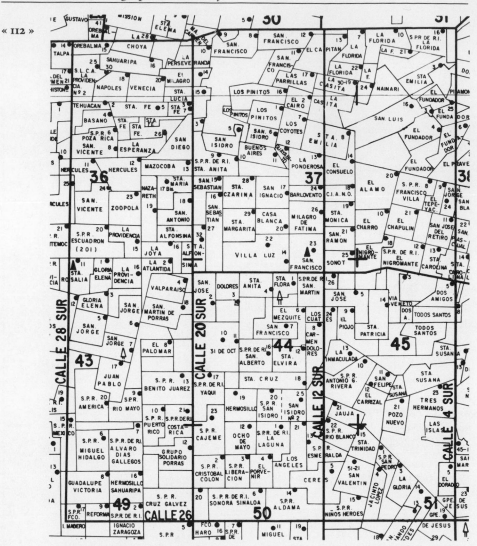

Property divisions, central part of Costa de Hermosillo, 1980s. (Source: Plano General, Secretaría de Agricultura y Recursos Hidráulicos, Distrito de Riego no. 51, Costa de Hermosillo, Feb. de 1984, 1:200,000.)

FIG. 31

will cause the abandonment of some holdings and a trend toward the cultivation of crops that require less water than the normal plantings of wheat and cotton.

Early on, the state government completed a rectangular grid of farm roads (many now paved) throughout the irrigated area to facilitate the harvesting and marketing of crops. Unlike the regular block survey system «113» found in the Río Yaqui delta, properties in the Costa de Hermosillo are irregular in shape and area (fig. 31). Each of the larger holdings contains one or more wells, from which water is distributed to fields via a complex arrangement of small canals that today are lined with cement or plastic sheeting to conserve water.

During the late 1970s a new farming area irrigated with well water (seventy-eight wells in 1989) was opened up in a small desert basin called Sahuaral, immediately south of the delta, but considered a part of the Costa de Hermosillo district.[66] As early as the 1950s wells farther south were drilled in Guaymas Valley (northeast of the port), which for a time was a small but thriving cotton and wheat area. However, because of serious overexploitation of the aquifer underlying the basin, saltwater infiltration has occurred and many wells and farms in the area have now been abandoned. Today almost all of the fresh water pumped in the valley is piped to the port of Guaymas, leaving only brackish water for irrigation.[67]

Located in the heart of the Altar Desert of northwestern Sonora, the Concepción-Magdalena delta has become a well developed irrigated farming area only since the late 1960s. Pima Alto Indians may have used incipient canal irrigation in the vicinity of Caborca even in pre-Conquest times, a practice continued by Jesuit missionaries in the colonial period; and during the late eighteenth century, haciendas, such as that of Arituaba, used water from the lower Magdalena to irrigate food crops sold to miners in northern Sonora. The Mexican government first became involved in irrigation within the area by constructing a small reservoir on the Río Altar, a northern tributary of the Rio Concepción (1947–1952). But, as in the Costa de Hermosillo, most water comes from the aquifers tapped by deep wells drilled to depths of fifty to ninety meters. In 1989 several hundred wells were in operation in the Caborca irrigation district, supplying water for some one hundred thousand hectares in wheat, cotton, and vineyards.[68] Although the water table has dropped steadily since pumping began twenty-five years ago, the cultivated area continues to expand, some fields reaching the coastal dunes that border the Gulf of California.

The Delta of the Colorado River

The extreme northwestern part of Sonora covers only a small portion of the Río Colorado delta, most of which lies within Baja California at the northern end of the Gulf of California. Since the late 1930s most of the delta has been converted from a desert wasteland to a productive irrigated

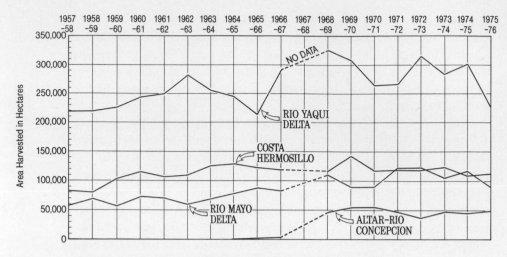

Area harvested in major irrigated districts of western Sonora, 1957–1976.

FIG. 32

farming district (El Valle de Mexicali), fed by river water via canals from both the Mexican and the American sides. The lower part of the river's main channel marks the border between Sonora and the state of Baja California Norte. In 1944, owing to a westward shift of the main channel, Sonora's share of the irrigable land was increased, so that today the state claims approximately one-fifth of the cultivated portion of the delta on its eastern flank.[69] The agricultural production of this land, of course, adds to Sonora's total farm income, but soil salinity is common, causing the abandonment of many farms. Marketing of the crops (mainly cotton and wheat) takes place in the Sonoran international border city of San Luis Río Colorado, which since the 1930s has grown from a small ranch settlement to an urban center of more than one hundred thousand inhabitants. Although connected with Hermosillo by rail and paved highway, its economic and social ties are more with northern Baja California and the United States than with western Sonora.

Figure 32 indicates the relative size of area harvested in each of the four major irrigated districts of western Sonora for the 1957 to 1976 period. The Yaqui delta's area far exceeds that of the other three, but its annual totals fluctuate greatly, from 210,000 to 320,000 hectares—probably a reflection of (1) changes in water levels of the reservoirs along the Yaqui River, caused by the sequence of wet and dry years and (2) problems of soil salinity. The harvested areas of the Mayo delta vary less, possibly because

of the river's relatively small catchment basin. Even less variable are the
areas of the Costa de Hermosillo and the Altar–Río Concepción (Caborca) district, in both of which water is obtained from wells. That a
systematic program of well drilling in the Caborca area was not begun
until the mid-1960s is seen in the sudden increase of the harvested area in
the latter part of the decade; by the late 1980s Caborca's irrigated area had
expanded to more than one hundred thousand hectares, nearly equal to
that of the Costa de Hermosillo.

The Crop Complex

The decision of farmers to cultivate particular crops in the irrigated districts of western Sonora has been determined by various factors: governmental pressure to produce food (for example, wheat and oilseeds) for the
growing Mexican population or to concentrate on crops (for example,
cotton and winter vegetables) that could be sold abroad to increase national foreign exchange; in view of water shortages, many farmers in
northwestern Sonora are now shifting to perennial crops (for example,
fruit and nut trees and vineyards), which require less irrigation than most
of the annuals. In any case, the tendency to shift crop patterns may be due
more to changing price and market opportunities than to bureaucratic
pressure.[70]

In all of the irrigated districts of western Sonora wheat is the dominant
crop, occupying from 30 to 50 percent of areas planted annually.[71] As it is
elsewhere in Mexico, in Sonora wheat is a winter crop, for it is adversely
affected by rust disease during summer rains. Especially after the development of new, high-yielding varieties during the "green revolution," the
increased use of fertilizers and the application of insecticides, after 1950
caused wheat production to soar in the irrigated districts of western Sonora.[72] There, the yield per hectare is twice the world average, and today
the state produces 40 percent of the nation's wheat crop, supplanting
the Bajío in the central part of the country as Mexico's "breadbasket"
(fig. 33).[73]

Cotton has been the main summer crop in the deltas of western Sonora
since the federal government intervened in irrigation during the 1940s.
Cotton once played a key role in Mexican agriculture, for it was a leading
earner of foreign exchange, three-quarters of the annual crop being exported to Japan. But because of competition with other producing countries and a drop in the world price, since the mid-1970s the acreage of
cotton in Mexico has decreased sharply. For example the area planted to

Irrigated wheat field ready for harvest, Costa de Hermosillo. The high-tension line along the farm road (left) supplies power for operating pumps that lift water from deep wells.

FIG. 33

cotton in western Sonora decreased from more than 135,000 hectares in 1973 to fewer than 50,000 in 1988.[74] Formerly gathered by hand with cheap migratory labor, today in the Yaqui delta, the Costa de Hermosillo, and the Caborca district, the crop is now harvested by mechanical pickers, as in the United States.

Oilseeds, especially sesame (*Sesamum indicum*) and safflower (*Carthamus tincorius*), both ancient cultigens of the Old World, have been cultivated on a large scale in many irrigated areas of northern Mexico since the 1950s. The cooking oils refined from these two plants have become common in Mexican kitchens throughout the country. In western Sonora production is especially important in the Mayo and Yaqui deltas.[75] Oil is also obtained by processing cottonseed after ginning the fiber, and some farmers grow flax on a small scale for linseed oil.[76] Even the native desert plant, jojoba, is being cultivated on an experimental basis; the seeds can be processed to obtain an oil used to lubricate fine machinery.

The basic Mexican native foods, maize and beans, are relatively minor crops in the Sonoran deltas and are grown mainly for local consumption. However, since the 1970s the cultivation of soybeans and grain sorghum (a mixture is used to produce animal feed) and the lowly chick-pea (garbanzo) (for human food) is of increasing importance in the crop complex of irrigated farming in western Sonora.[77]

Although early attempts were made in the Mayo and Yaqui deltas to produce winter vegetables for the United States market, this industry has never been significant in Sonora. As late as the 1940s some tomatoes, fresh peas, and bell peppers were grown for export in the Guaymas, Yaqui, and Mayo areas, but because of occasional winter frosts farmers suffered considerable losses. Thus, the industry was retained in the frost-free coastal lowlands of Sinaloa, where today the irrigated farms from Culiacán southward to Nayarit today form the major winter vegetable region of Mexico.[78]

One of the most interesting recent developments in crop diversification in the Costa de Hermosillo, Guaymas, and Caborca irrigation districts revolves around the tendency of farmers to shift from seasonal cotton and grains to perennials such as fruit trees and alfalfa. As indicated earlier, the problem of decreasing water supply from wells has been a major cause of this change; furthermore, the national market for fruit has been favorable. Citrus, peaches, grapes, and olives are the major fruits that have been planted, beginning mainly in the 1970s. Pecans are also important. Of these, grapes are by far the most significant in terms of hectares planted and financial return. Over a period of ten years (1972–1982) the area devoted to vineyards in western Sonora increased from fewer than two thousand to thirty-one thousand hectares, making the state the leading producer of grapes in Mexico.[79] Nearly two-thirds of the hectarage is now in the Caborca area, the rest in the Costa de Hermosillo and the Valle de Guaymas. Production, based almost entirely on the Thompson Seedless variety, consists of table grapes (a quarter of which are exported to the United States), raisins (mainly in the Caborca area, owing to its aridity and low year-round humidity), and the manufacture of wine and brandy (local wineries dominated by the Domecq and Vergel companies).[80] In the Costa de Hermosillo the area planted to citrus (mainly oranges) has doubled since 1984. Olives, however, are limited largely to the Caborca district,[81] where the close association of citrus orchards, vineyards, olive groves, and date palms presents an agricultural landscape reminiscent of southeastern Spain (fig. 34).

Despite the agricultural production of the Sonoran and other irrigated

Olive grove (right) and vineyard (left), Caborca irrigation district, northwestern Sonora.
The lateral canal (center) carries water from a pump in far distance to irrigate both fields.

FIG. 34

districts of northern Mexico, the country still cannot feed itself and is unlikely to be able do so in the near future because of its growing population and the limited amount of land suitable for farming. Today Mexico imports large quantities of wheat and wheat flour, maize, oilseeds, soybeans, and sorghum from the United States.[82]

Mechanization, the Labor Force, and Farm Animals

As in most areas of irrigated agriculture in northern Mexico, almost from its inception in the late 1920s large-scale commercial agriculture in western Sonora has been mechanized.[83] These modern practices essentially reduced the need for a large permanent labor force.

Before the recent decline of cotton and the introduction of mechanical pickers, however, large numbers of migratory workers were needed to harvest the crop. For instance, in the Yaqui delta during the 1950s and 1960s between sixty thousand and eighty thousand migrants from various parts of Mexico arrived each year in the late summer months to pick cotton,[84] just as Mexican seasonal laborers (*braceros*) entered the United States to bring in crops. A similar influx of migrants seeking work must have occurred also in the Mayo delta and the Costa de Hermosillo, where cotton was the major summer harvest. It is estimated that perhaps half of the migrants stayed on, hoping to find work either in the fields or in the cities, thus adding to the growing population of the "New Sonora".[85] The recent growth of the fruit industry in the Hermosillo and Caborca irrigation districts has led to the need for seasonal labor, especially for the grape harvest, but not on the scale that the cotton harvest once required.

Again, in part by reason of mechanization, the need for farm animals is lacking in Sonora's new irrigated agriculture. Horses, mules, and burros, so common in the eastern part of the state, as elsewhere in Mexico, are conspicuous by their absence. A few dairies located on the edge of the irrigated fields produce milk for the local urban centers. But more obvious are large chicken farms (*plantas avícolas*), where Leghorn hens and pullets are confined to elongated sheds mainly for egg production (fig. 35). These have been part of the rural scene in western Sonora since the 1960s, as they have been in other parts of the country. In addition, the 1980s saw the introduction of specialized hog farms (*granjas porcinas*), where the animals are kept in large sheds similar to those used for poultry and practically force-fed to produce pork and lard. In both cases, the farms are owned and operated by nonfarming specialists who depend on the locally processed soybean-sorghum feeds to produce eggs and meat mainly for the Mexico City market. Both chicken and hog farms usually occupy nonirrigable lots within or on the edge of the cultivated areas along main roads. Today the commercial chicken population of western Sonora exceeds 13 million (89 percent of the state total); hogs number nearly 350,000 (87 percent of the state total), most of which are found in the specialized animal "factories".[86]

Land Tenure

Initially characterized by private ownership, land tenure in the irrigation districts of western Sonora was not seriously affected by national agrarian reform policies until the mid-1930s. In October of 1937 the government, by order of President Lázaro Cárdenas, expropriated seventeen thousand hectares of privately held irrigated land in the Yaqui delta for establishing collective ejidos to be given to local landless peasants who had been clamoring for the means of making a living.[87] Although given only use rights of the land, the ejido members (*ejidatarios*) continued the commercial

A chicken farm (granja avícola) *near the main highway and adjacent to an irrigated area, south of Ciudad Obregón, Valle del Yaqui.*

FIG. 35

farming of wheat and cotton, rented machinery from neighboring property owners, and divided proceeds from crop sales among the group members. Later, through expropriations and expansion of irrigation, more ejidos, both collective and individual types, were established in the Yaqui delta, so that by the 1960s they comprised more than a third of the irrigated land, excluding the Yaqui pueblo holdings along the river.[88] In the Mayo delta, much of which was already in Indian hands, ejidos were also established, and today the land is nearly equally divided between ejido and private holdings.[89]

In contrast, in the more recently developed Costa de Hermosillo and Caborca districts private ownership prevails. Despite the agrarian law limiting private holdings of irrigated land to one hundred hectares, some individuals in the Costa de Hermosillo were able to operate illegally farms in excess of one thousand hectares by grouping adjacent parcels held by relatives.[90]

To encourage settlement of the Costa by less-affluent private farmers, in 1949 the government opened a large area of national land (*baldíos*) to

colonists, who were sold small parcels on easy terms. The colonists (*co-*
lonos) were grouped into units of ten families, each unit occupying two
hundred hectares and each family within the unit owning twenty hectares;
moreover, the government helped the *colonos* to rent farm machinery to
cultivate their parcels.[91] The system was apparently devised mainly to help
cover the cost of drilling deep wells to obtain irrigation water, since a
single family could hardly bear the expense alone. In fact, the reliance on
groundwater for irrigation may have discouraged the formation of ejidos
in both the Hermosillo and the Caborca areas, owing to the high cost of
drilling wells. After its establishment in the Costa de Hermosillo in 1949,
the *colono* program soon spread to the Guaymas, Caborca, and Yaqui districts, where holdings range from ten to one hundred hectares in size; by
1975 the number of *colonos* in the Hermosillo area alone numbered 765.[92]

It is widely recognized that private holdings have proved to be the most
efficient and productive units in the highly mechanized irrigated agriculture of northwestern Mexico. Most of the individual *ejidatarios* tend to
grow subsistence crops, such as maize, whereas the owners of private
properties produce crops for the marketplace. Because of lack of proper
mechanical equipment and the difficulties of internal administration, even
the collective ejidos cannot compete with private farmers, who take better
care of their land and crops than do the *ejidatarios*. In the Yaqui and Mayo
deltas many *ejidatarios* find it more profitable to rent their parcels illegally
to the private farmers and to hire themselves out as farmhands than to
work their own land.[93]

Sonora's Fishing Industry

About the same time that government-sponsored irrigated districts were
being established in western Sonora, another significant activity—commercial fishing—was evolving along the Gulf coast of the state. This development was to increase further the economy of New Sonora and attract
additional population to the arid coast.

Due in part to its high phytoplankton productivity, the Gulf of California teems with many species of finfish, crustaceans, and mollusks.[94] Moreover, since the post-Pleistocene rise of sea level, the numerous marsh- and
mangrove-bordered lagoons and estuaries that line the Gulf coast of Sonora and Sinaloa have served as nurseries for juvenile forms of crustaceans,
principally shrimp, and several kinds of marine fishes that spend most of
their adult life in open waters. As previously described, the wealth of sea-
food was utilized in aboriginal times by Seri fishers, who inhabited the

Partial view of Puerto Peñasco, northwestern Sonora. Part of the shrimp fleet and freezing plants line the dock, upper center.

FIG. 36

desert coast of central Sonora, and by the Yaqui, Mayo, and other Cáhita-speaking Indians of southern Sonora and northern Sinaloa. Today shrimp and sardines compose by far the bulk of the commercial catch along that coast.

As early as the 1920s a few Mexican shrimp boats were operating out of Guaymas,[95] which later would become the major fisheries center of Sonora. In the 1930s the Japanese, with permission of the Mexican government, began exploiting shrimp beds off Guaymas. During their voyage through the Gulf of California in 1940, novelist John Steinbeck and naturalist E. F. Ricketts observed the Japanese operations, noting that some six trawlers were actually dredging the sea bottom, scooping up tons of shrimp and other marine life, but saving only the former for processing; the bycatch, which consisted of dead or dying fish, was dumped overboard.[96] Alarmed by this wasteful process, Mexican authorities soon banned Japanese fleets from the Gulf. Nonetheless, the Japanese experience stimulated Mexican involvement in commercial fishing and led to the

Part of the fishing fleet docked at Yávaros. Freezing plants for shrimp and a sardine cannery are seen on the right.

FIG. 37

establishment of three major fishing centers (Guaymas, Puerto Peñasco, and Yávaros) as well as several smaller fishing communities and camps along the Sonoran Coast (figs. 27, 36, 37, table 1).[97]

Today fleets of trawlers work out of the major centers, taking the small shrimp (*Penaeus californiensis*) that inhabit the deep hypersaline offshore waters of the northern part of the Gulf.[98] But like most shrimpers the world over, the Mexicans still ensnare all types of marine life in their trawls and keep the shrimp and discard most of the bycatch, which could be processed and utilized either for commercial fertilizer or as a valuable food source.[99] Onshore the cleaned shrimp are deep frozen and shipped to market; nearly three-quarters of the catch is exported to the United States in refrigerated trailer trucks.[100]

The smaller fishing communities are found near or on the shores of coastal lagoons and estuaries, where the large shrimp species *P. stylirostris* abounds. This species spends a much longer time than *P. californiensis* in shallow water before migrating into Gulf waters as adults.[101] In the la-

The Three Major Fishing Centers of Sonora, 1989

	Shrimp Trawlers	Sardine Boats	Fish Canneries	Freezing Plants	Cooperatives (and Members)
Guaymas	366	38	5	12	38 (4,055)
Puerto Peñasco	212	0	0	4	19 (1,748)
Yávaros	19	11	1	2	No data

Source: Sonora 1990a: 154–55, 157–58.

TABLE 1

goons fishers use throw nets (*atarrayas*) and weirs (*tapos*) built across narrow mouths of tidal inlets to entrap shrimp and other marine life; both methods have aboriginal origins.

Since the 1950s most of the lagoonal fishers have organized into cooperatives, a system that permits them to acquire such tools as fiberglass boats powered by outboard motors and other equipment for their trade. However, they account for only 15 to 20 percent of the Sonoran shrimp catch.[102] Varying in population from fewer than fifty to more than one thousand persons, the small fishing communities contain many families that came to Sonora from other parts of Mexico as migrant farmhands, possibly attracted by the cotton boom of the 1960s, and that subsequently turned to fishing for a livelihood. Other communities, such as Guásimas and Líliba (Lobos), south of Guaymas, are inhabited mainly by Yaqui Indians who claim exclusive fishing rights in those lagoons thought to be within their traditional territory.[103]

Combining the catches of the trawlers in open water and from the lagoons and estuaries, today Sonora produces about one-fifth of Mexico's shrimp harvest.[104] This crustacean alone makes up 80 percent of the total value of the state's fisheries. A third source of shrimp production—the raising of the animal in indoor hatcheries, a form of aquiculture—was begun at Puerto Peñasco in the 1970s and is becoming increasingly significant.[105]

In terms of tonnage Sonora's catch of a second marine species—sardines—is six times that of shrimp, but in value it is far less.[106] The term "sardine" may encompass a variety of small and juvenile fish. In the Gulf of California a herringlike genus (*Oposthpterus*, spp.) and an anchovy (*Anchoa mudeoloides*) perhaps form the bulk of the "sardine" catch. Both species run in enormous schools and are easily trawled with sardine boats out

A recreational beach at Huatabampito near Yávaros, Valle del Mayo. It is lined with sun shelters (palapas) for weekend bathers from Navojoa and other towns within the Valle del Mayo, southern Sonora.

FIG. 38

of Guaymas and Yávaros. Sonora is Mexico's premier sardine-producing state; in 1989 it accounted for 60 percent of the nation's catch, a third of which is canned in Guaymas for national consumption.[107] An undetermined amount, together with a large tonnage of the bycatch, is processed for poultry feed and fishmeal used as fertilizer.

Coastal Recreation

The recent development of the Sonoran coast is seen not only in the rise of the fishing industry, but also in the use of the beaches for recreation and tourism (fig. 27). On weekends and holidays, especially during the summer, city and town folk, both the affluent and the poor, flock to the many sandy beaches that line the coast. The people of Hermosillo drive cars or ride buses to Bahía Kino, one hundred kilometers to the west; those from Ciudad Obregón and the towns in the Yaqui delta enjoy the

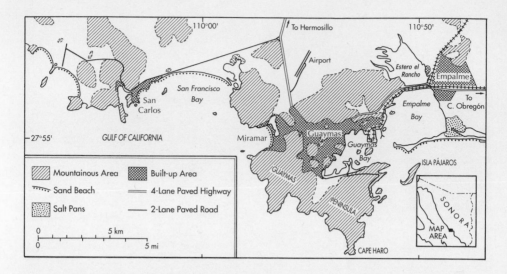

Port city of Guaymas and adjacent tourist areas.

FIG. 39

nearby beaches of Los Médanos and Los Halcones; and the folk of Na-vojoa and the Mayo delta frequent the resort of Huatabampito near Yávaros.

Most of the weekend beaches contain small stands that dispense beer and soft drinks, and back from the surf are long lines of palm-thatched sun shelters (*palapas*) for the bathers (fig. 38). But many of the more wealthy families have built substantial winter homes along some of the beaches, the largest and most ostentatious example being at Nuevo Bahía Kino on the coast west of Hermosillo. A solidly occupied eight-kilometer stretch of elaborate Spanish-type houses occupies both sides of a paved highway along the beach north of the fishing settlement. Developed by an affluent landowner after 1965,[108] Nuevo Bahía Kino now contains homes belonging not only to rich families from Hermosillo, but also to some North Americans from California and Arizona, who enjoy their seaside villas during the pleasant Sonoran winters.

In terms of international tourism Sonora's most renowned resort center lies in the vicinity of Guaymas. A mountainous coast interrupted by small bays and sandy crescent beaches within a desert climate has given this area a unique physical setting for the development of seaside resorts (fig. 39). San Carlos, west of the port of Guaymas, attracts tourists from the United States, Canada, and Europe as well as from various parts of Mexico.

Partial view of Hermosillo, state capital of Sonora and a growing industrial center of 400,000.

FIG. 40

Population and Industry in Western Sonora, 1985

City (Municipio)	Population	Number of factories	Number of employees
Hermosillo	406,963	918	11,021
Ciudad Obregón (Cajeme)	302,687	704	9,139
Navojoa	125,224	220	2,212
Guaymas	125,505	217	3,375
San Luis Río Colorado	105,560	175	2,028
Caborca	58,796	123	831

Source: Sonora 1990*a*: 22–24; Sonora 1990*b*: 123–124, 125–127.

TABLE 2

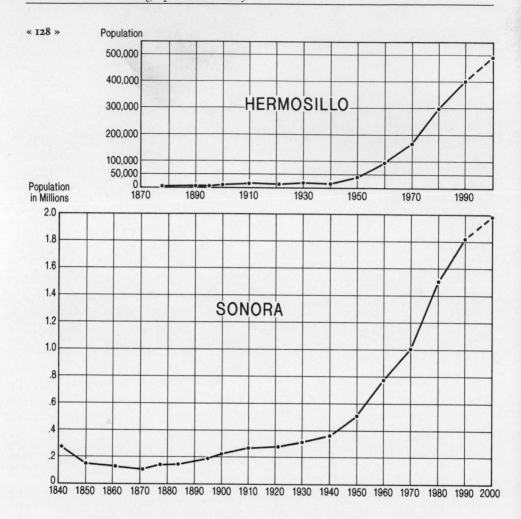

Population growth, state of Sonora and city of Hermosillo, nineteenth and twentieth centuries.

FIG. 41

Poorly occupied during the hot summers, the hotels, apartments, and trailer parks are usually filled in winter by visitors who come to enjoy the pleasantly dry air, the desert scenery, and especially the sport fishing. In 1989 touristic activities in the Guaymas/San Carlos area afforded jobs for more than six thousand people, a greater number than in any other Sonoran urban center.[109]

Maquiladora *Plants in Sonora, 1986*

	Population	Number of plants	Number of employees
Nogales	90,853	51	15,252
Agua Prieta	40,749	26	6,325

Source: Sonora 1990*a*: 21, 23; South 1990.

TABLE 3

Population, Urbanization, and Industrialism

Western Sonora has seen a spectacular population growth, stemming mainly from the development of irrigated agriculture during the last half century. The eastern part of the state has been relatively stagnant. Much of western Sonora's growth derives from the influx of farmers and migrant workers attracted by the possibility of high crop yields and permanent or seasonal labor in the fields. Equally important, in each of the main irrigation districts a large urban center has evolved, largely to attend the various activities related to processing units, canneries, wineries, and other crop-processing plants that have formed a growing industrial sector within each of the cities and along main roads within the irrigated areas. And this phenomenon has drawn more and more people into the urban service sectors.

Again, the mechanization of agriculture has attracted agencies, many of them branches of North American companies, specializing in selling and repairing modern farm machinery. Similarly, the widespread use of commercial fertilizers and pesticides has induced the establishment of chemical plants and sales outlets in the cities. In the districts dependent on groundwater for irrigation (Hermosillo and Caborca) well-drilling concerns have found a lucrative business. Moreover, the export of agricultural products abroad and to other parts of Mexico has necessitated the services of shipping companies, brokers, and banks. Finally, beginning in the 1930s, the involvement of the federal government in water control gave rise to regulatory agencies, such as the Secretaría de Recursos Hidráulicos, whose regional offices make up a good part of the local bureaucracy in Sonoran cities.[110]

Such industries and commercial outlets are typical of all the larger cities

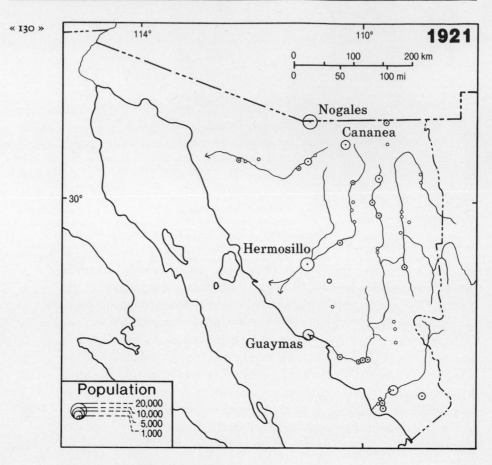

Population distribution, state of Sonora, 1921.

FIG. 42

of western Sonora: the state capital, within the Costa de Hermosillo; Ciudad Obregón in the Valle del Yaqui; Navojoa in the Valle del Mayo; and Caborca in the delta of the Asunción-Magdalena river system (fig. 40). In the Colorado River delta the city of San Luis Río Colorado serves the larger Baja California portion of the Mexicali irrigation district perhaps as much or more than it does the smaller Sonoran part.

Although most of the industrial development of these cities is based on local agriculture, cheap labor and a ready market have drawn other types of manufacturing, such as the Ford Motor Company's auto assembly plant established at Hermosillo in the mid-1980s [111] and the new brewery, a branch of Cervecería Cuauhtémoc, in Navojoa. (The port of Guaymas is

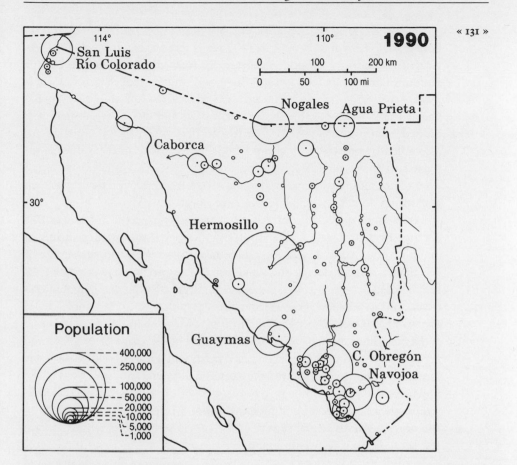

Population distribution, state of Sonora, 1990.

FIG. 43

the only large western Sonoran city whose industry is based chiefly on fish processing, rather than on agriculture.) Responding to financial support from the state government, various Mexican interests have lately established industrial parks, open to all kinds of manufacturing, on the outskirts of all of the large cities.[112] Practically all industrial plants in Sonora are powered by electricity generated by local thermoelectric plants or by hydroelectric stations constructed on the Yaqui and Mayo rivers.[113] Table 2 summarizes the status of urban centers and industry in western Sonora. In 1985, 77 percent of Sonora's manufacturing establishments and 56 percent of its industrial workers were concentrated in the six largest cities in the western part of the state.[114]

« 132 »   Since the late 1960s two northern Sonoran cities on the international border—Nogales and Agua Prieta—have entered the industrial arena. As part of its National Frontier Program, in 1965 Mexico agreed to permit United States companies to form "in-bond" transnational factories (*maquiladoras*) in the northern border cities from the Gulf of Mexico to the Pacific (table 3). Under this arrangement parts and raw materials from the United States are taken duty-free to the border towns, where they are assembled or processed by cheap Mexican labor; the final product is returned to the United States for marketing. In both Nogales and Agua Prieta light industrial items such as electronic and electrical devices as well as clothing dominate the labor-intensive operations.[115]

Following a period of relatively slow growth during the first half of this century, from 1950 to 1990 the population of Sonora increased at an annual rate between 3.5 and 4 percent, a pace faster than the national average of less than 3 percent during the same period. Approximately 85 percent of the increase occurred in western Sonora, largely within the irrigated districts and associated cities. Hermosillo's population, for example, from 1950 to 1990 increased between 4 and 6 percent per annum (fig. 41).

During the same period, except for the border cities of Nogales and Agua Prieta, the population of eastern Sonora grew slowly. There, the farmers and ranchers inhabited the same small towns and hamlets as in the nineteenth century. Although by 1990 two of the towns had increased in size to twenty thousand and fifteen thousand respectively (Magdalena in the north, Alamos in the south), most of the settlements ranged from a few score persons to five thousand.

Figures 42 and 43 depict Sonora's population (places with one thousand or more persons) in 1921 and 1990. In 1921 the three towns (Hermosillo, Guaymas, and Nogales) along the north-south economic axis were the largest urban places in the state, and clusters of small settlements were present in the Yaqui and Mayo deltas. Eastern Sonora, although it contained nearly 70 percent of the state's population, could boast only a few small settlements of one thousand or more persons. Seventy years later a complete reversal in population had occurred, occasioned by the rapid growth of cities and towns in the western part of the state, all closely associated with the development of commercial farming under irrigation. This reversal is illustrated in figure 44, which shows population trends in the two sections of the state, 1900–1990. At the beginning of this century, only slightly more than 30 percent of Sonora's population inhabited the arid west, whereas nearly 70 percent still lived in the east, mainly in the

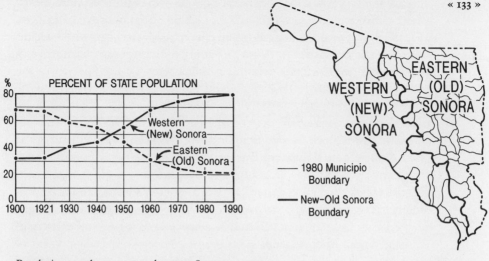

Population trends, western and eastern Sonora, 1900–1990.

FIG. 44

narrow river valleys of La Serrana. From 1900 on, the percentage steadily declined in the east, but rose in the west, so that by 1945 the proportion between the two sections had equalized. Today 80 percent live in the west.

Nearly 80 percent of western Sonora's population is now urban (that is, with a population of five thousand or more), according to the definition of the term "urban" used by some demographers for developing countries. In contrast, excluding the border cities of Nogales and Agua Prieta, where in-bond (*maquiladora*) industries are dominant, the urban population of eastern, or "Old" Sonora, totals only 15 percent; the remainder is rural or in mining.

As indicated earlier, the population growth in western Sonora was occasioned in large part by the influx of migrant and permanent farmhands who came mainly from the neighboring states of Sinaloa, Baja California, and Chihuahua; others hailed from densely peopled areas of central Mexico—Jalisco, Michoacán, México, and the Federal District.[116] Eventually, with increased mechanization and decreased demand for seasonal labor on the farms, many of the immigrants returned to their home areas, but large numbers drifted into Sonoran cities to seek employment in industry, construction, and services that required little expertise. Others settled in small

« 134 » agricultural towns and hamlets, especially in the Yaqui and Mayo deltas, where occasional jobs as farmhands and service employees are often available. Such settlements, including the ejidos, form clusters in both deltas, in contrast to the dispersed farmsteads and processing plants characteristic of the Costa de Hermosillo and the Caborca districts.

Today, in numbers of people, agricultural development, urbanization, and industrialization, western and eastern Sonora each has its distinct way of life: one is urban and industrial and supported by a mechanized commercial agricultural base; the other is conservatively rural, underlain by a largely subsistence-type economy and inhabited by people in many ways steeped in their colonial past. Thus, Sonora is a single political entity with a dual geographical personality.

A *Jesuit* Memoria

Following is an example of an annual request for supplies from Mexico City, sent in by Father Felipe Segesser, Jesuit missionary stationed at Tecoripa. It is dated 1736 (AHH, Temporalidades 2016, exped. 2).

  6 *arrobas* [150 pounds] of fine chocolate and
  2 *arrobas* [50 pounds] of ordinary quality
  1 package [*tercio*] of sugar
  1 pound of saffron
  1 pound of cinnamon
  2 pounds of [black] pepper
  ½ pound of nails
  6 pounds of quinine [*quinquibre*]
100 yards of woolen sackcloth [*sayal*]
 50 yards of fine woolen cloth
 50 yards of flannel [*bayeta*]
  1 *arroba* [25 pounds] refined beeswax
 12 good-quality axes
  1 cloth cassock
  6 white sieves and 6 black ones
  ½ dozen colored cloths
  2 pounds of snuff
  ½ dozen metal L-plates to secure some boxes
  4 steel cooking plates [*comales de acero*]
  1 ream of paper
  1 dozen China plates; another dozen of wood
  2 bed blankets
  1 new missal containing all the new feast days

    6 dozen sailor's knives

    2 pounds of green, yellow, and red chomite [coarse woolen cloth for making women's skirts]

    ½ dozen bouquets of church [artificial?] flowers

    1 small book entitled *Christian Prophet*

In his relation on Sonora, Father Segesser tried to justify his large use of chocolate by stating that it was the custom in Sonora, as in other parts of Mexico, to serve hot chocolate to both religious and lay Spanish guests at his mission (Treutlein 1945:160). Other luxury foodstuffs and household articles are listed. Cloth and clothing were habitually given to Indian workers and their families living near the mission, probably in order to retain their loyalty.

Property of the Mission of Sahuaripa, 1735

(AHH, TEMPORALIDADES 277, EXPED. 73)

Livestock

Large animals [*ganado mayor*]
277 mares distributed in various herds [*manadas*] pastured near Sahuaripa;
in San Mateo 64 mares obtained through trade in Tepari for as many
burros and a large jack [*garañón*]
62 young mules, large and small
95 colts
5 tamed male mules
47 tamed female mules
32 untamed female mules [*mulas serranas*] at San Mateo
74 tamed horses
44 oxen
Livestock with nursing young: 148 head [*ganado chichiguo*]
Small animals [*ganado menor*]
370 sheep, large and small
14 hogs

Cultivated Land

1 orchard with fruit trees and sugarcane
1 field of sugarcane consisting of 158 rows and a field of chile, both located
downriver [from Sahuaripa?]
1 field at Xunupo, irrigated, ready for harvest [crops not specified],
1 *fanega*, 10 *almudes* [ca. 9½ acres]
1 field, irrigated [crops not specified], 2 *fanegas*, 10 *almudes* [ca. 11 acres]
plus 3 *almudes* [2½ acres] of beans, with irrigation canals ready for use.

Church Lands along the Lower Bavispe Valley, 1790

A document of 1790 lists the size of church lands (*labores de comunidad*) in several pueblos along the Río Moctezuma and the lower Río Bavispe well after the Jesuit expulsion. The large size of the fields given in this list probably was not typical for missions of an earlier period (AF-BNM 35/772).

Opotu

32 *fanegas* of wheat field [20.4 hectares, or 51 acres]

Bacadéhuachi

2 *labores de comunidad*
  (1) size of field, 1,200 *varas* by 200 *varas* [yards] [20 hectares, or 50 acres]
  (2) size of field: 680 *varas* by 220 *varas* [12.4 hectares, or 31 acres]
1 garden [*huerta*], 540 by 140 *varas* [6 hectares, or 15.6 acres]

Batuco

2 *labores de comunidad*
  (1) 700 *varas* by 300 *varas* [17.6 hectares, or 44 acres]
  (2) 400 *varas* by 200 *varas* [6.4 hectares, or 16 acres]

1 *labor de comunidad*, 750 *varas* by 300 *varas* [18.6 hectares, or 46.5 acres]

Guachinera

2 *labores de comunidad* in wheat and maize
   (1) 702 *varas* by 264 *varas* [14.8 hectares, or 37 acres]
   (2) 540 *varas* by 264 *varas* [11.6 hectares, or 29 acres]

Sellos, *1684 and 1714*

Samples of *sellos* (orders) for obtaining Indians to work in Sonoran mines under the *repartimiento* system (AHH, Temporalidades 325, exped. 87).
Sample 1.
With this document I order and command the [Indian] governors and *topiles* of Cucurpe, Opodepe, and Tuapa that every 15 days, 20 Indian *tapisques* be brought to this valley [Bacanuche] for the development of the mines in which His Majesty has much interest for the royal income; that the Indians are to be well paid and well fed; and do not contradict this order on pain of being punished . . . ; of the said 20 Indians 6 are to be taken from Tuapa, 6 from Opodepe, and 8 from Cucurpe. . . .

Valley of Bacanuchi, 1 February 1684, I hereby swear as *juez receptor* before witnesses,

Jerónimo García, *juez receptor*

Sample 2.
Governors, *alcaldes*, and *topiles* of the pueblo of Onabas: You will hand over to the carrier of this *sello*, 6 *tapisques* for work in the mines [of Río Chico]; they will be well treated and paid; and do not do anything contrary to this order on pain of punishment for disobeying the command of the royal court [*real justicia*].

Your vice-*alcalde mayor*,

Río Chico, December 7, 1714.

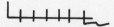

[*Note*: The crude ideogram that appears at the end of the *sellos* apparently was for the benefit of the native governors and *topiles*, who were probably illiterate.]

Monthly Statistics on the Mining Camp of San Francisco de Asís, 1805–1809   (AF-BNM 37/824)

	Gold Produced in *Marcos* (1 *marco* = 8 oz.)	Population Estimates (Workers)	Number of Stores	Number of Traders (*Rescata-dores*)	Number of Field Kitchens (*Cocinas*)
1805					
Jan.	250	4,500	32	80	80
Feb.	196	4,500	36	—	173
Mar.	100	—	—	—	—
Apr.	125	600	15	10	—
May	75	600	—	—	—
June	60	—	—	—	—
July	70	—	—	—	—
Aug.	91	3,500	—	—	—
Sep.	313	5,000	21	50	50
Oct.	272	5,000	—	—	—
Nov.	251	4,000	23	60	40
Dec.	211	4,000	—	—	—
1806					
Jan.	256	4,000	23	56	40
Feb.	283	4,500	23	58	46
Mar.	347	5,000	27	61	42
Apr.	253	3,400	23	53	41
May	211	1,500	20	37	28
June	132	600	10	17	6
July	87	800	—	—	—
Aug.	238	2,500	—	—	—
Sep.	210	5,000	—	—	—

Oct.	193	3,000	12	20	16
Nov.	155	—	—	—	—
Dec.	168	—	12	40	13
1807					
Jan.	—	—	—	—	—
Feb.	165	1,000	14	36	14
Mar.	120	1,500	12	38	16
Apr.	168	4,400	17	50	20
May	—	—	—	—	—
June	815	400	14	16	8
July	74	1,000	14	12	8
Aug.	96	2,000	15	30	10
Sep.	219	3,000	18	4	18
Oct.	—	—	—	—	—
Nov.	129	2,000	16	48	20
Dec.	95	2,000	—	—	—
1808					
Jan.	—	—	—	—	—
Feb.	83	1,500	15	46	0
Mar.	107	1,600	14	40	0
Apr.	79	3,000	16	46	0
May	74	2,000	15	40	0
June	70	1,500	14	28	0
July	42	1,400	14	30	0
Aug.	46	3,000	—	—	—
Sep.	54	1,800	14	35	0
Oct.	144	2,500	14	40	0
Nov.	140	2,600	16	50	0
Dec.	150	2,600	16	45	0
1809					
Jan.	139	2,600	16	50	0
Feb.	125	2,600	16	45	0
Mar.	140	2,600	16	50	0
Apr.	98	1,600	14	34	0
May	—	—	—	—	—
June	80	1,000	16	16	0

Notes

Preface

1. H. Clifford Darby, "On the Relation of Geography and History," *Transactions and Papers of the Institute of British Geographers* 19 (1953): 1–12.

2. The *Tableau* was published in 1903 as vol. 1, pt. 1 of *Histoire de la France*, ed. E. Lavisse (Paris: Librairie Hachette). The section entitled "La Personnalité géographique de la France" illustrates Vidal's concept of the term. This section was republished in 1941 in England with a preface by British geographer H. J. Fleure (Manchester: Manchester University Press; London: Hachette).

3. Dunbar, 1974; Sauer, 1941; Emyr Estyn Evans, *The Personality of Ireland: Habitat, Heritage and History* (new enlarged ed., Belfast: Blackstaff Press, 1981).

4. Vidal de la Blache, "Les Genres de vie dans la géographie humaine," *Annales de Géographie*, vol. 20 (1911): 193–213, 289–304.

5. Max Sorre, "La Notion de genre de vie et sa valeur actuelle," *Annales de Géographie*, vol. 57 (1948): 97–108, 193–204. English translation ("The Concept of Genre de Vie") in Philip L. Wagner and Marvin W. Mikesell (eds.), *Readings in Cultural Geography*. (Chicago and London: University of Chicago Press, 1962, pp. 399–415).

1. Physical and Biotic Aspects of Sonora

1. Merrill 1905; King 1939.

2. Hastings and Turner 1965; Hastings and Humphrey 1969.

3. Turnage and Mallery 1941; Hastings and Humphrey 1969.

4. According to Santamaría (1959:498), the word *equipata* derives from a Tarahumara Indian expression. In central Mexico light winter rains are termed *cabañuelas*, a word that refers to the first twelve days of January (ibid.:163).

5. Blake 1935. The most recent tropical storm to affect the Sonoran area occurred June 6–8, 1990.

6. Kalstrom 1952; Rosendal 1963.

7. Damon et al. 1962; Livingston and Damon 1968; Anderson, et al. 1970; Cserna 1970.

8. Waibel 1928.

9. Turnage and Mallery 1941; Hastings and Turner 1965; Fontana 1974:496.

10. McGee 1897; Waibel 1928.

11. Lumholtz and Dracopoli 1912:505; Fontana 1974:501.

12. Felger 1980:87.

13. Shreve and Wiggins 1951, I:25.

14. Shreve 1934; Shreve and Wiggins 1951, I:24.

15. Standley 1920, XXIII; pt. 1, pp. 72, 74; Gentry 1942:86; Schabel 1962.

16. Gentry 1942:27–30; Wiseman 1980:145–146; D. Brown 1982:101–105.

17. Bahre and Bradbury 1980.

18. Shreve and Wiggins 1951, I:frontispiece, map 2.

19. S. White 1948:230–238.

20. D. Brown 1982:59–65; Doolittle 1988:14.

21. Dice 1939:118.

22. Goldman and Moore 1945; Leopold 1952; Pennington 1980, I:152.

23. Goldman and Moore 1945; Leopold 1952.

24. Pennington 1980, I:192; Riley 1987:63.

25. Bahre 1967:41.

2. *The Aboriginal Cultures of Sonora*

1. Pailes 1978, 1980; Doolittle 1984*a*, 1984*b*, 1988.

2. Doolittle 1984*a*.

3. Doolittle 1984*b*, 1988:46–47.

4. Braniff 1978, 1984.

5. Sauer and Brand 1931:105, 106.

6. Amsden 1928:45, 47.

7. Riley 1987:39, 89–90; Doolittle 1984*a*, 1988:58–60.

8. Doolittle 1988:61.

9. Doolittle 1984*b*:246; Riley 1987:57.

10. Riley 1987:85–86; Pailes 1980:24.

11. Doolittle 1984*a*.

12. Riley 1987:39, 325.

13. Pérez de Ribas 1985, II:9, 84; Fabila 1940:46; Spicer 1980:5.

14. Pérez de Ribas 1985, II:10, 84.

15. Ibid.:84.

16. Spicer 1980:10.

17. Castetter and Bell 1942:1.

18. Sauer and Brand 1931:119.

19. Pennington 1980, I:2.

20. Ibid.:145.

21. Sauer 1935*a*:5; Pennington 1980, I:36.

22. Castetter and Bell 1942:1.

23. According to Sauer (1935a:5), perhaps thirty thousand.

24. Fontana 1974:516.

25. Castetter and Bell 1942:40, 168; Fontana 1974:518–519.

26. Castetter and Bell 1942:48.

27. Bolton 1919:I, 50–51, n. 193; Mange 1926:271; Castetter and Bell 1942:156; Fontana 1974:520.

28. Mange 1926:217.

29. Sauer 1935*a*:5.

30. Nabhan and Felger 1978:3, 5.

31. Nabhan et al. 1979:173.

32. Sauer and Brand 1931.

33. A. Johnson 1963; Harlem 1964.

34. Di Peso and Matson 1965; Bahre 1967:36, 40.

35. Felger and Moser 1970, 1976.

36. For example, see the April 1970 issue of the journal *The Kiva*, devoted entirely to Seri studies.

37. Lockwood 1938; Mails 1974; Haskell 1987.

38. Haskell 1987:71.

39. Forbes 1957, 1959, 1960.

40. Haskell 1987:83.

3. Spanish Settlement of Sonora: The Missions

1. Literature on Jesuit mission history in Sonora is abundant and varied. Among the important works used to construct the summary presented here are John F. Bannon, *The Mission Frontier in Sonora, 1620–1687*, which outlines the Jesuit activities in southern and eastern Sonora; the mission history in the Pimería Alta narrated in Herbert E. Bolton's *The Rim of Christendom*; Frank G. Lockwood's *With Padre Kino on the Trail*; and John A. Donohue's *After Kino: Jesuit Missions in Northwestern New Spain, 1711–1767*. Charles W. Polzer deals with the operational methodology and philosophical basis of Jesuit activity in Sonora in *Rules and Precepts of the Jesuit Missions in Northwestern New Spain*. Basic to all secondary works on early Jesuit activity in Sinaloa and Sonora is Padre Andrés Pérez de Ribas, *Historia de los triunfos de Nuestra Santa Fé . . .* (1 vol.), first published in 1645 with subsequent Spanish editions in 1944 (3 vols.) and 1985 (2 vols.). A condensed version in English has appeared: *My Life among the Savage Nations of New Spain*, trans. Tomás Antonio Robertson (Los Angeles: Ward Ritchie Press, 1968).

2. Pérez de Ribas 1985, II:14; Roca 1967:327; Crumrine 1977:16.

3. Crumrine 1977:19.

4. Polzer 1976:8.

5. Polzer 1976:49.

6. Polzer 1976:17, 63.

7. Spicer 1980:16, 17.

8. Roca 1967:315.

9. See Polzer 1976:34 for a map showing the areas covered by rectorates in Sonora and northern Sinaloa.

10. Bannon 1955:30, 48.

11. Ibid.:46.

12. Ibid.:50.

13. Ibid.:86–88; Polzer 1972.

14. Polzer 1972.

15. Bannon 1955:129.

16. Lockwood 1934:67.

17. Ibid.:72; Bolton 1936:500ff.

18. Roca 1967:89–90.

19. Donohue 1969:114ff.

20. Pfefferkorn 1949:275–226.

21. Polzer 1976:5.

22. Donohue 1969:19–31.

23. Spicer 1962:56–57, idem 1980:59–61; Crumrine 1977:17; J. Johnson 1950:38ff.

24. Polzer 1976:63.

25. According to Father Segesser, stationed at the mission of Tecoripa in 1738, "the King and the superiors of the Order have directed that Indians learn also in Spanish the Christian doctrine and necessary prayers" (Treutlein 1945:160).

26. Pfefferkorn 1949:246, 269.

27. J. Johnson, 1950:43–44; Spicer 1980.

28. Treutlein 1945:151; Pfefferkorn 1949:275.

29. Treutlein 1945:151.

30. Pfefferkorn 1949:275.

31. Treutlein 1945:151.

32. Archivo Histórico de Hacienda, Mexico City (hereafter cited as AHH), Temporalidades 278, exped. 24 (1673); Archivo General de Indias, Seville (hereafter cited as AGI), Patronato 232 (1673).

33. Pfefferkorn 1949:204.

34. Ibid.:94.

35. Treutlein 1965:140.

36. AHH, Temporalidades 279, var. expeds.

37. Treutlein 1945:185; idem 1965:140; Pfefferkorn 1949:98–100.

38. Polzer 1972:171–172, 183–184.

39. AHH, Temporalidades 279, exped. 29.

40. Treutlein 1965:140; Nentvig 1957:672.

41. Spicer 1980:30, 63, 120.

42. Archivo General de la Nación, Mexico City (hereafter cited as AGN), Jesuitas II, exped. 92.

43. Archivo Franciscano, Biblioteca Nacional de México, (hereafter cited as AF-BNM) 35/775; Ocaranza 1937:I:282.

44. AF-BNM 35/775; Ocaranza 1937:I:286.

45. Pfefferkorn 1949:245–246.

46. Treutlein 1945:152.

47. Nentvig 1957:672.

48. For an example list of the livestock and agricultural property of a Sonoran mission in the early eighteenth century, see Appendices B and C.

4. Spanish Settlement of Sonora: The Mines and Ranches

1. Merrill 1908a:802–803.

2. AHH, Temporalidades 278, exped. 24. The court proceedings were held at the mining settlement of San Miguel Arcángel.

3. Navarro García 1967:252.

4. Polzer 1972:254.

5. Bancroft (North Mexican States) 1884, I:232; Almada 1952:560.

6. AHH, Temporalidades 1126, exped. 1.

7. Ibid.

8. Polzer 1972:255.

9. Navarro García 1967:67–68; Mange 1926:340–346.

10. Almada 1952:562.

11. These included Juan de Oliva, Miguel de Casanova, Laureano Bascón de Prado, and Juan Mungía Villeta (ibid.:103, 149, 492, 531).

12. Archivo Municipal del Parral (hereafter cited as AMP), 1649 Administración, exped. 13; Almada 1952:588–590.

13. Almada 1952:561.

14. AGN, Provincias Internas 29, exped. 6 (1750); AGN, Historia 16 (unnumbered exped.) 1750, f. 245; AGI, Audiencia de Guadalajara 137 (unnumbered exped.) 1757, f. 261ᵛ.

15. AMP, 1658 Administración, exped. 4.

16. AGI, Escribanía 400-A, cuad. 1 (1681); AMP, 1682 Administración, exped. 7.

17. AMP, 1718 Administración, exped. 2.

18. Aldama 1952:499.

19. AGN, Provincias Internas 30, exped. 8 (1689). The *real* was described as "one of the most opulent of this Province," but because of an uprising of the Suma, Jacome, and Jano Indians, "four rich mining families have left, and others are about to do so" (f. 238).

20. AHH, Temporalidades 17, exped. 42. "The *real* of Nacazori, one of the richest mining centers of this province, had a very large Spanish population until the year 1755. . . . The people have left because of the assaults, thievery, and murders that the enemy [the Apache] make every week".

21. Pradeau and Burrus n.d.:188; AMP 1685 Administración, exped. 102; AGI, Audiencia de Guadalajara 152, unnumbered exped., 1695, 1698; 154, unnumbered exped., 1691; Biblioteca Nacional de México, MS Div. (hereafter cited as BNM-MS) 489 (1705), f. 407.

22. AGN, Historia 316, exped. 11; AMP 1666 Adminstración, unnumbered exped.; Navarro García 1967:38; Polzer 1972:145.

23. AMP, 1688 Minas, exped. 139.

24. Nentvig 1980:118, n. 14; Pennington 1980, I:75.

25. Navarro García 1967:69. The author based his information on data in AGI, Escribanía 400-C, exped. 2. He assumed that each militiaman was a head of a family of seven, giving a total Spanish population of 1,302 for Sonora in 1684. Bolton and Marshall (1920:240) give an estimate for the year 1678: "In Sonora the people of Spanish or mixed blood numbered about 500 families."

26. AGI, Escribanía 397-A, f. 122; Navarro García 1967:38.

27. Aldama 1952:543.

28. AF-BNM 36/819.

29. Gerhard 1982:266.

30. AGN, Jesuitas II, exped. 19; AHH, Temporalidades 17, exped. 42; 278, AHH,

« **148** »   Temporalidades exped. 41; AMP, 1723 Administración exped. 107; AF-BNM 12/200 bis (1724); Ocaranza 1937–1939 II : 39.

31. AF-BNM 35/781, 35/784.

32. AMP, 1718 Administración, exped. 2.

33. AGI, Audiencia de Guadalajara 185, unnumbered exped., 1738; Tamarón y Romeral 1937 : 265–66. The AGI documents reported that the largest piece weighed over one hundred arrobas (twenty-five hundred pounds), an obvious exaggeration. See also Bancroft 1884, I:256, n. 15; Merrill 1906; Donohue 1969 : 83.

34. In some instances the *alcaldes mayores* inspecting mines and ranches in Sonora appeared to have been more interested in collecting *mordidas*, or fees from miners and stockmen, than in enforcing regulations. For example, Francisco Cuervo de Valdés, *alcalde mayor* of Sonora (1681–1685), on one of his inspection trips was charging each mine owner a fee of fifty pesos to receive a satisfactory report on the status of his mines and furnaces in operation (AMP, 1682 Administración, exped. 7).

35. Ibid.; 7; AGI, Escribanía 400-B, cuaderno 6.

36. AMP, 1677 Administración, exped. 104; AGI, Escribanía 400-B, cuad. 6. According to the *Recopilación de leyes de los reynos de las Indias*, Libro VI, título 15, ley 12, the use of forced Indian labor to drain mines was prohibited; this ruling was blatantly ignored in Sonora.

37. AMP, 1707 Administración, exped. 12.

38. In 1737 Father Segesser (Treutlein 1945 : 186) commented: "the miners prefer using their powder in the mines to using it for hunting," a statement that suggests that blasting was being practiced in Sonora at that time. This is quite early, for the technique was not used in central Mexico until the beginning of the eighteenth century. By 1790 blasting powder was listed as one of the imports into Sonora that was subject to the *alcabala*, or sales tax, indicating that it was widely used in the mines at that time (AGN, Provincias Internas 259, exped. 7, ff. 236, 243ᵛ).

39. AHH, Temporalidades 279, exped. 28 (undated, probably 1690). This document mentions the use of a small mill with only two stamps.

40. AGI, Audiencia de Guadalajara 145, unnumbered exped., 1689; AHH, Temporalidades 282, exped. 11 (1690).

41. AMP, 1707 Administración, exped. 12; AMP, 1728 Civil, exped. 123; AMP, 1720 Civil, exped. 108.

42. AGI, Escribanía 400-B, cuad. 6 (1686).

43. AMP, 1720 Civil, exped. 108. In 1791 the intendant of Sonora reported to the viceroy in Mexico City that 450 mule-powered *arrastres* were in operation at the revived *real* of Aigamé (AF-BNM 35/784).

44. AHH, Temporalidades 325, exped. 71 (1673); AGI, Audiencia de Guadalajara 472 (1772). During colonial times the Yaqui exploited the coastal salines bordering the Yaqui River delta, hauling salt by mule train into the mines and other Spanish settlements in the interior (Ocaranza 1937–1939, I : 287).

45. Biblioteca Pública del Estado, Guadalajara (MS Div.), Civil (1700–1719), leg. 2, exped. 4. The term *tepusque* derives from the Nahuatl *tepuzqui*, meaning copper. Small pieces of copper, called tepusques, were used as a medium of exchange during the early Spanish period in Mexico (Santamaría 1959 : 1036), and the expression persisted in colonial Sonora for irregularly shaped bits of silver used as coins for trading.

46. AGI, Audiencia de Guadalajara 135; Burrus 1971:524, 525.

47. AMP, 1663 Administración, exped. 104; AMP, 1664 Administración, exped. 13; AMP, 1669 Administración, exped. 4; AMP, 1677 Administración, exped. 133; AMP, 1688 Administración, exped. 139.

48. Navarro García 1967:38–39. Using a ten-year (1675–1684) series of statistics on silver production in Parral and Sonora (AGI, Audiencia de Guadalajara 18), Navarro García (1967:39) showed that only 22.5 percent of the combined production of the two places originated in Sonora.

49. Bloomer 1909:699; Pearce 1911:688.

50. Stewart 1955:21.

51. Aldama 1952:735.

52. Bolton 1936:240.

53. AGI, Audiencia de Guadalajara 19, unnumbered exped., 1687; AGI, Escribanía 391-A, unnumbered exped., 1689. These two documents detail the actions of the *alcalde mayor* of Sinaloa, Gen. Domingo de Terán de los Ríos, relating to the establishment of a *real de minas* to serve the new discoveries in the Sierra de Alamos. Soon after mining began in the sierra, the *alcalde* issued an order forcing all miners to settle in a mining center called Guadalupe some five leagues (twenty-five kilometers) north of the puesto of Los Alamos. The *real* of Guadalupe was near the left bank of the Río Mayo, where, according to the *alcalde*, large water-powered stamp mills could be constructed, where salt was readily available from the coastal salines, and where a church and attendant priests could serve the inhabitants. Apparently, opposition to the order was so great that it was rescinded, leaving the miners, merchants, and Indian workers free to form an urban center at the much-closer puesto of Los Alamos, despite serious drawbacks to the site.

54. Stewart 1955:70.

55. Acosta 1946:59.

56. Tamarón y Romeral 1937:240.

57. Bancroft Library (MS), Province of Sonora (Libro de la Real Hacienda, Real de Alamos, 1770–1776).

58. Stewart 1955:73.

59. AHH, Temporalidades 278, exped. 36 (1671); AHH, Temporalidades 974, exped. 1 (1750); AGN, Provincias Internas 29, exped. 6 (1750). Several lists of merchandise carried into Sonora, most of them dating from the last quarter of the seventeenth century, detail the types of cloth, clothing, and other articles; such lists are found in various documents in the Parral Archives, for example, AMP, 1673 Administración, exped. 118.

60. AMP, 1682 Administración, exped. 7.

61. AGI, Escribanía 400-A, cuad. 1.

62. Examples of manifests of silver bullion reduced from ores by smelting are found in various documents in the Parral Archives, for example, AMP, 1669 Administración, exped. 4.

63. AHH, Temporalidades 278, exped. 50 (1677); Tamarón y Romeral 1937:294; Dodge 1973:40–43.

64. AGI, Audiencia de Guadalajara 145 (informe del gobernador de Nueva Vizcaya, nov. de 1684).

65. Alegre 1841–1842, II:11.

66. Navarro García 1967:319. Around 1650, at the village of Tomóchic on the Yécora

trail, Tarahumara Indians attacked a Sonoran mule train carrying salt, tallow, and silver to Parral. (*Documentos para la historia de México*, ser. 4, III : 179).

67. AGI, Audiencia de Guadalajara 13, unnumbered exped. 1674; AGI, Audiencia de Guadalajara 154, unnumbered exped. 1691.

68. Acosta 1946 : 54.

69. AGN, Provincias Internas 29, exped. 6 (1750); AGN, Provincias Internas 81, exped. 1 (1771).

70. AMP, 1649 Administración, exped. 13.

71. AMP, 1685 Administración, exped. 102; AMP, 1678 Administración, exped. 104 (*visita* of 1677).

72. By law and custom a *sitio* for cattle and horses (*ganado mayor*) was a square block of pastureland approximately 5,000 Spanish yards (*varas*), or 4.2 kilometers, on the side, comprising an area of about 1,755 hectares. However, the shape of the tract varied. Examples of early land grants to miners in Sonora include the following: Salvador de Valenzuela claimed that his title to a ranch (presumably a *sitio*) in the Cedros Valley of Ostímuri was granted in 1645 (AMP 1685 Administración, exped. 102). Around 1655 a *sitio* was granted to Pedro Ruiz de Andaya, a miner in the *real* of San Pedro Reyes. The grant was located in the mountains some 7 leagues from the site of San Juan Bautista, west of the Río Moctezuma. After the death of Ruiz, the land passed into the hands of relatives; the title was reaffirmed in the local court in 1675 (AMP, 1675 Terrenos y Sitios, exped. 20).

73. AHH, Temporalidades 325, exped. 33 (1713); Donohue 1969 : 21.

74. AGN, Jesuitas I, exped. 11 (1715); AHH, Temporalidades 325, exped. 24 (1713).

75. AMP, 1723 Administración, exped. 103.

76. AGI, Audiencia de Guadalajara 145 (governor of Nueva Vizcaya to the viceroy, 20 Jan. 1673).

77. Ibid. (governor of Nueva Vizcaya to the king, 4 Mar. 1674).

78. AMP, 1715 Administración, exped. 136.

79. AGN, Historia 72.

5. Mine and Mission Relations in Colonial Sonora

1. Pfefferkorn 1949 : 273; Treutlein 1939 : 289–290, 297.

2. Polzer 1972 : 146–47.

3. AGI Patronato 232, ramo 1 (folios 46–127).

4. Transcript of the hearings is recorded in AHH, Temporalidades 278, expeds. 24, 36; AHH, Temporalidades 325, exped. 69; AGN, Historia 316, exped. 12.

5. Pfefferkorn 1949 : 274.

6. AHH, Temporalidades 278, exped. 3.

7. AGN, Historia 316, exped. 14; Treutlein 1939:297; Navarro García 1967 : 217–223.

8. AGN, Historia 316, exped. 12 (f. 456^v); AGI, Patronato 232, ramo 1 (ff. 203–206).

9. This precept was repeated in decisions of Jesuit superiors as late as 1747 (Polzer 1976 : 17, 31, 115, 116, 122).

10. AGI, Patronato 232, ramo 1, ff. 249, 249^v; Polzer 1972 : 182.

11. AHH, Temporalidades 279, exped. 28.

12. AHH, Temporalidades 278, exped. 36.

13. AHH, Temporalidades 325, exped. 69.

14. AGN, Provincias Internas 152, exped. 1; AF-BNM 35/775; Ocaranza 1937–1939, « 151 »
I : 287, 298.

15. Simpson, 1938 : 11.

16. Ibid. : 13–15. Various regulations regarding the repartimiento in Mexico are found in *Recopilación de leyes de los reynos de las Indias*, libro 6, títulos 12 and 15.

17. Polzer 1972 : 156.

18. AGI, Escribanía 400-C, cuad. 9, f. 24ᵛ.

19. AGN, Indios 40, exped. 29, ff. 57ᵛ-70ᵛ. This document was published in part in Simpson 1938 : 154–57.

20. Bolton 1936 : 233–235.

21. Colección Mateu (Barcelona), Informe de Lorenzo José García, 1744; Tamarón y Romeral 1937 : 246–247; Spicer 1980 : 126, 220, 321; Hu-DeHart 1981 : 41, 52–53.

22. AGN, Provincias Internas 30, exped. 7 (1689).

23. AGN, Provincias Internas 69, exped. 2, ff. 73, 75 (1761).

24. AGI, Audiencia de Guadalajara 505, unnumbered exped. In 1705 the Jesuit missionary stationed at San Francisco de Batuco on the middle Río Yaqui stated: "During the last fourteen years the Indians of my mission have discovered more than sixty mines, but the ores are nothing more than of copper, arsenic, and antimony; there are no silver ores. . . . Four years ago many mines near the *real* of Aigamé were discovered by Pima Indians." (BNM-MS 489, f. 405). The Batuqueños (Pima Bajo of the Batuco villages) were noted for their interest in mining.

25. Navarro García 1965 : 398.

26. AGI, Patronato 232, ramo 1, ff. 56–57.

27. Ibid., f. 57ᵛ.

28. Mexico, Censo general de población, 1900.

29. Spicer, 1962, 1980.

30. AGI, Patronato 232, ramo 1, f. 82ᵛ.

31. AHH, Temporalidades 278, exped. 41, ff. 6ᵛ-7ᵛ.

32. Ibid., f. 7ᵛ.

33. Bancroft 1884, I : 572–574, n. 46.

34. AF-BNM 36/806 (1797).

35. AGI, Audiencia de Guadalajara 284, exped. 291.

36. W. B. Stevens Collection no. 68 (1765), Benson Latin American Collection, University of Texas, Austin; Tamarón y Romeral (1937) : 298–310.

37. Reff 1991 : 103, 108, 114.

38. Sauer 1935*a*:11; Reff 1991 : 98, 132.

39. Crumrine 1977 : 18.

40. Polzer 1972 : 172; Reff 1991 : 142, 160, 173.

41. Fabry 1743 : 5; Gerhard 1972 : 23.

42. Reff 1991 : 179.

43. Pfefferkorn 1949 : 217–219, 264.

44. AHH, Temporalidades 325, exped. 8.

6. Indian Depredations in Sonora

1. Thrapp 1967 : 358–359.

2. Tamarón y Romeral 1937 : 240, 254; Bancroft 1884, I : 574.

3. Mendizábal 1930 : 119; Forbes, 1960; Naylor and Polzer 1986 : 483.

4. AGN, Provincias Internas 30. exped. 5; AGN, Jesuitas II, exped. 10 (1686).

5. Haskell 1987 : 83–84.

6. Galaviz de Capdevielle 1968 : 5.

7. AGI, Audiencia de Guadalajara 152, unnumbered exped., *alcalde mayor* of Sonora to governor of Nueva Vizcaya, June 1690.

8. Treutlein 1965 : 179–180; Pfefferkorn 1949 : 145.

9. Fehrenbach 1974 : 86.

10. AGI, Audiencia de Guadalajara 253, ff. 54–54ᵛ (1780).

11. Griffen 1988*b* : 6–7; 151–159.

12. Nentvig, 1957–1958 : 672.

13. Bartlett 1852 : 100.

14. Mendizábal 1930 : 120.

15. Treutlein 1945 : 153.

16. Mendizábal 1930 : 120.

17. Bartlett 1852 : 100.

18. Pfefferkorn 1949 : 212.

19. AHH, Temporalidades 278, exped. 20.

20. Donohue 1969 : 54.

21. For example, in AHH, Temporalidades 17, exped. 42, there appears a list of over one hundred atrocities that the Apache, Seri, and Pima Alto committed in Sonora between 1755 and 1760.

22. Galaviz de Capdevielle 1968 : 5.

23. AGI, Audiencia de Guadalajara 152, unnumbered exped.; AGI, Audiencia de Guadalajara 154, unnumbered exped., ff. 132ᵛ-143ᵛ.

24. AGI, Audiencia de Guadalajara 135, unnumbered exped., Mange to Governor Huidobro, 8 July 1735.

25. Bartlett 1852 : 95.

26. One of the first militias in Sonora was organized under the command of Capt. Pedro Perea in 1645 to quell an uprising of Pima Alto Indians in the northern part of the province.

27. AF-BNM 34/738.

28. As late as the 1850s, Bartlett (1852 : 97) commented on the military skill of the Opata Indians (at that time almost completely Hispanicized) in fighting the Apache. The Opatas would commonly rout the enemy in a pitched battle, whereas the Apache usually defeated Mexican troops. On one occasion the Opatas rescued 860 head of stolen livestock.

29. AGN, Provincias Internas 47, exped. 1, ff. 31, 33, 37, 41.

30. AGN, Provincias Internas 48, exped. 3.

31. AGN, Provincias Internas 82, exped. 1, f. 445ᵛ, 448.

32. Spicer 1962 : 234.

33. AGI, Audiencia de Guadalajara 152, unnumbered exped.; AF-BNM 12/200 bis (1704), ff. 114, 116.

34. Hernández 1971 : 33; Doc. para hist. de Méx. 3ʳᵈ ser. I, pt. 2 : 605, 607, 608; Shull 1968.

35. AGN, Provincias Internas 259, exped. 7, ff. 256–256ᵛ.

36. Rippy 1919 : 388–389; idem 1926 : 77, 80; Park 1962 : 325 ff., Griffen 1988*a* : 14–15.

37. Griffen 1985:142.

38. AGI, Audiencia de Guadalajara 514, unnumbered exped. In 1775 the military roster of the presidio of Fronteras was as follows:

 1 captain—European [peninsular Spaniard?]
 1 lieutenant—Spanish [*criollo*]
 1 sublieutenant—mestizo.
 1 sergeant—Spanish [*criollo*]
 2 corporals—Spanish [*criollo*]
25 soldiers—Spanish [*criollo*]
 5 soldiers—mulattoes
 9 soldiers—mestizos
10 "exploradores"—Opata Indians
55 total

Additional population of this presidio is composed of 45 families, all *gente de razón* [whites or mixed bloods], no one among them being of any Indian nation nearby; but all are illiterate.

39. AF-BNM 34/738.

40. AGI, Audiencia de Guadalajara 514, unnumbered exped. (1773). In 1804 the presidio of Altar had a population of 945, Tucson, 1,019, Tubac, 84 soldiers and families, 20 families of Indians, 8 families of *gente de razón* (AF-BNM 36/819).

41. AGI, Audiencia de Guadalajara 518, exped. 33 (1784). In 1783, for example, Don Esteban Gach, merchant of Arispe, obtained a contract to provide necessities to the presidios of Santa Cruz (Terrenate), Tucson, Altar, and Pitic for a period of five years, beginning in 1784.

42. Blackmore 1891:194.

43. AGI, Audiencia de Guadalajara 253 (*informe*, Croix to Gálvez, 23 Jan. 1780).

44. Rippy 1926:77.

45. Fehrenbach 1974:208, 210; Richardson 1933:193–210.

46. Sauer 1935b:3; Donohue 1969:154.

47. Sauer 1935*b*:3.

7. The Sonoran Gold Craze

1. Until the mid-eighteenth century most of the gold mined in colonial Mexico was a by-product of silver ores. In central Mexico the silver mines of San Luis Potosí and Guanajuato were among the major producers of gold, but in addition there were smaller mining centers, such as El Oro, near Tlalpujahua in Michoacán and San Francisco del Oro in Chihuahua, based on the exploitation of mainly gold ores.

2. AF-BNM 35/764, f. 4; Ocaranza 1937–1939, II:48.

3. AGN, Provincias Internas 245A, exped. 13, f. 74^v.

4. Ibid., f. 77; AGN, Historia 16, f. 206; Tamarón y Romeral 1937:243.

5. Ocaranza 1937–1939, II:178.

6. Biblioteca Nacional de Madrid (Manuscript Division, hereafter referred to as BNM-Madrid), MS 19266.

« 154 »   7. AGN, Provincias Internas 82, exped. 1, f. 146ᵛ; AGN, Provincias Internas 47, exped. 1, f. 33.

8. AF-BNM 35/764, f. 4.

9. AGI, Audiencia de Guadalajara 519, exped. 101. During the 1760s gold placers were worked at several points along the upper and middle portions of the Río Sonora and its tributary arroyos: Chinapa, Arizpe, Motepore, and near Nacámeri, west of Ures.

10. Ocaranza 1937–1939, II : 179.

11. AGN, Misiones 14 (1784), ff. 222ᵛ—232; Ocaranza 1937–1939, II : 90; Pfefferkorn 1949 : 92.

12. Antonio de los Reyes 1938 : 304.

13. AGN, Misiones 14 (1772), f. 61ᵛ.

14. Ibid. f.62.

15. Pfefferkorn 1949 : 92.

16. AGN, Provincias Internas 86, exped. 2, f. 149.

17. AF-BNM 35/781, f. 1ᵛ.

18. AGN, Provincias Internas 86, exped. 2, f. 177; AF-BNM 35/781,f. 1ᵛ.

19. AGN, Provincias Internas 86, exped 2, f. 177.

20. Ibid.

21. Ibid.

22. AGI, Audiencia de Guadalajara 416, exped. 65, pt. 1 (1771); AGI, Audiencia de Guadalajara 416, exped. 73, pt. 1, f. 97ᵛ (1771); AGN, Provincias Internas 81, exped. 1, no. 135, ff. 16–16ᵛ (1771); Robertson 1777, II:328–329; Navarro García 1964:205–206, 266.

23. Bonillas 1911 : 156–157; Webber 1935 : 1–14.

24. AGI, Audiencia de Guadalajara 416, exped. 73, f. 97ᵛ; AGN, Provincias Internas 81, exped. 18, f. 256.

25. Richards 1911 : 797–798.

26. AGN, Provincias Internas 81, exped. 18. f. 198ᵛ (1772); AGN, Provincias Internas 232, exped. 1, ff. 167–167ᵛ(1772).

27. AGN, Provincias Internas 81, exped. 18, f. 385 (1772); AGN, Provincias Internas 232, exped. 1, f. 167ᵛ.

28. AGI, Audiencia de Guadalajara 416, exped. 65; AGN, Provincias Internas 81, exped. 1 (no. 139), f. 23 (1771).

29. AGI, Audiencia de Guadalajara 512, unnumbered exped. (1771).

30. AGN, Provincias Internas 81, exped. 18, f. 206. Other estimates of the population ranged as high as seven thousand in 1772 (Navarro García 1964 : 254).

31. AGN, Provincias Internas 81, exped. 18, f. 536; AGN, Provincias Internas 226, exped. 2, ff. 468–469.

32. AGN, Provincias Internas 81, exped. 18, f. 233ᵛ.

33. AGN, Provincias Internas 93, unnumbered exped., f. 230; AGI, Audiencia de Guadalajara 512, unnumbered exped., Corbalán to Croix, Aug. 2, 1771; AGN, Provincias Internas 81, exped. 18, f. 199ᵛ.

34. AGN, Provincias Internas 81, exped. 18.

35. Ibid., f. 198; AGI, Audiencia de Guadalajara 513; AGN, Provincias Internas 247, exped. 16, ff. 351ᵛ, 356ᵛ.

36. AGN, Provincias Internas 93, exped. 1, f. 128; AGI, Audiencia de Guadalajara 416, exped. 68, no. 4.

37. AGN, Provincias Internas 93, ff. 334–335ᵛ; AGN, Provincias Internas 81, exped. 18, f. 236ᵛ.

38. AGI, Audiencia de Guadalajara 416, exped. 68, no. 4; AGN, Provincias Internas 81, exped. 1, ff. 43–43ᵛ, 222.

39. AGN, Provincias Internas 232, ff. 209–211 (1772).

40. AGN, Provincias Internas 81, exped. 18, f. 192ᵛ (1772).

41. Ibid., ff. 224–226. In September 1772, 100,000 pesos worth of gold was packed in fifty boxes, loaded on thirty mules, and taken by military escort to the *caja real* in Chihuahua via the trail passing by the presidios of Terrenate and Janos, and through Púlpito Pass.

42. AGN, Provincias Internas 245-A, exped. 23, f. 192.

43. AGN, Provincias Internas 246, exped. 21, f. 337ᵛ.

44. AGN, Provincias Internas 247, exped. 16, ff. 361–366ᵛ.

45. AGI, Audiencia de Guadalajara 267 (1779), 253 (1780), 519 (1784).

46. AF-BNM 41/944, f. 1.

47. Bancroft Library, Pinart Papers. Ser. 1, items 23, 24.

48. AGI, Audiencia de Guadalajara 296, no. 33 (1804); Velasco 1850:196; Navarro García 1965:4–5.

49. AGI, Audiencia de Guadalajara 296, no. 87; AF-BNM 37/824. The latter reference consists of a long series of official reports on the status of Cieneguilla and San Francisco de Asís from 1805 through 1809. See Appendix E.

50. AGI, Audiencia de Guadalajara 296, nos. 106, 108, 110; Navarro García 1965:6, n. 7.

51. Velasco 1850:201ff.

52. Waring 1897:257.

53. Mowry 1864:87; Hamilton 1884:231; Lumholtz 1912*b*:83; Wilson 1952:15, 18–26; Bancroft 1889:499–500; 579–580; Heikes and York 1913:254–260; Love 1974:19–23, 57.

54. Velasco 1850:289, 291; Guinn 1909–1910; Borthwick 1857:139, 235, 311.

55. Southworth 1905:217; Merrill 1908*b*:360; Waring 1897:257.

56. Archivo Histórico del Gobierno del Estado de Sonora (Hermosillo), ramo Minería, Expediente Distrito de Altar, ff. 136, 139–143, 145–146.

57. Merrill 1908*b*:361.

8. *The Growing Domination of Western Sonora*

1. Voss 1982:152, 156.

2. Mowry 1864:93.

3. Salazar 1902:325.

4. D'Olwer 1965:1078; Calderón 1965:593.

5. AGN, Provincias Internas 70, unnumbered exped., ff. 224ᵛ–228; AGN, Provincias Internas 93, f. 217; Aldama 1952:323, 325; Voss 1982:44.

6. Mowry 1864:93–94; Voss 1982:152.

7. Mowry 1864:49.

8. Aldama 1952:341.

9. AGI, Audiencia de Guadalajara 518, unnumbered exped., ff. 1–14; Ocaranza 1937–1939, II : 165. Aldama (1952 : 144) states that the *villa* of Pitic was founded and lots distributed to *vecinos* in 1772; the AGI document cited here definitely indicates that the correct date of the founding is 1780.

10. Mowry 1864 : 50; Aldama : 344.

11. Voss 1982 : 108.

12. *Diccionario Porrúa* 1964 : 1013.

13. S. Ramírez 1884 : 578–579.

14. Ulloa 1910 : 71–72, 82, 94, 125, 135.

15. S. Ramírez 1884 : 83; Hafer 1912 : 903–904.

16. Ulloa 1910 : 78–79; Bernstein 1964 : 44–45.

17. Waibel 1927 : 571; Hu-DeHart 1984 : 160.

18. Ulloa 1910 : 87.

19. Humboldt 1811, II : 328.

20. Dabdoub 1964 : 383.

21. Among the best accounts of Yaqui uprisings during the nineteenth century is Hu-DeHart 1984.

22. Dabdoub 1964 : 113; Voss 1982 : 150–151.

23. Dabdoub 1964 : 125, 139.

24. Pfeifer 1939 : 446; Dabdoub 1964 : 268–272; McGuire 1986 : 33.

25. In 1894 Walter S. Logan of New York, the president of the Sonora and Sinaloa Irrigation Company, published a booklet entitled "Yaqui, the Land of Sunshine and Health," extolling the merits of the Yaqui delta and its glowing agricultural future. But his prognostications were not realized until more than a half century later.

26. Fabila 1940 : 105; Dabdoub 1964 : 285; Hu-DeHart 1984 : 161; McGuire 1986 : 33.

27. Dabdoub 1964 : 283, 289.

28. The individual blocks are designated in sequence by numbers, those east of the prime meridian ending in even numbers, those west of it, in odd numbers. The numbers of the first east-west row of blocks south of the baseline begin with 100, the second row with 200, and so on (e.g., 101, 103, 105 west of the prime meridian; 102, 104, 106 east of the prime). At present, the southernmost row of blocks in the northern part of the Mayo delta is designated by the 2900 numbers (sheet G12B44, Villa Juárez [Sonora], Carta Topográfica, 1 : 50,000, Estados Unidos Mexicanos).

29. Dabdoub 1964 : 287–288.

30. Ibid. : 308–309; Spaulding 1974 : 42.

31. Waibel 1927 : 573; Pfeifer 1939 : 446.

32. Pfeifer 1939 : 447.

33. One of the most popular propaganda tracts issued by the Richardson company was a booklet entitled "The Yaqui River Valley," published about 1920. Much of the information contained therein appears to have been copied directly from the pamphlet published in 1895 by the Sonora and Sinaloa Irrigation Company.

34. Dabdoub 1964 : 314.

35. Dabdoub 1964 : 321–324; Cajeme 1989 : 199. In 1926 there were some fifty Germans among the farmers (most of the latter were North Americans) in the Yaqui delta (Waibel 1927 : 573).

36. Dabdoub 1964 : 314, 317. To gather precise information on river characteristics,

by 1920 the Richardsons had established some fifteen stations along the Yaqui to measure discharge as well as climatic data.

37. Bond 1935:209; Pfeifer 1939:446; Dabdoub 1964:318.

38. Dabdoub 1964:330.

39. Fabila 1940:58.

40. Pfeifer 1939:448.

41. Waibel 1927:574.

42. Dabdoub 1964:326–327.

43. Pfeifer 1939:438–442; Benassini 1942:93.

44. Waibel 1927:574. According to Waibel, since 1911 tomatoes had been shipped to the United States from the Fuerte area. In 1915, five hundred railroad cars, each containing 840 boxes of tomatoes reached the United States; during the 1925–1926 harvest period twenty-six hundred cars made the trip.

45. Orive Alba 1945:13–15; Gulhati 1958:8–9.

46. Orive Alba 1945:15.

47. Ibid.

48. Orive Alba 1946:9.

49. Mexico, informe presidencial 1948:131–132.

50. Loma 1963:18; Merriam 1957:431.

51. Spaulding 1974:44.

52. Vargas Alacántara 1959:21; idem 1960:102–103.

53. Anon. 1943:40; Merriam 1957:431.

54. Loma 1963:20; Anon. 1967:25.

55. In Sinaloa the federal government continued the development of the several deltas along the coast by constructing large reservoirs similar to those completed on the lower Yaqui and Mayo rivers in Sonora. By the late 1960s major dams had been finished on the Fuerte, Sinaloa, Culiacán (on the Tamazula and Humaya tributaries), and San Lorenzo rivers, with plans to continue work farther south. These projects greatly increased the irrigated areas of the coastal plain, where private enterprise had already led to modest developments. In the 1970s various schemes were proposed to construct interbasin canals for a more equitable distribution of river water between the deltas, including those of southern Sonora, but none of these ideas were carried to fruition (Cummings 1974:11).

56. Ríos 1960:27–28; Anon. 1961:71; Dozier 1963:569–571; Henderson 1965:304–305; Anon. 1971a:table, Mapa de areas afectadas por salinidad, pp. 73–74.

57. Thomson 1989:29–38.

58. Rowan 1962:57.

59. Jiménez Villalobos 1965:65; Busch et al. 1966:163; Matlock et al. 1966:166; Sanderson 1981:151; Hermosillo 1988:95.

60. Mexico, 1948.

61. Hermosillo 1988:187.

62. Rowan 1962:67.

63. Jardines Moreno 1976:8–9, 11.

64. Jiménez Villalobos 1965:66.

65. Ing. Roberto Biebrich Aguayo, jefe de riego y drenaje, Costa de Hermosillo, Secretaría de Agricultura y Recursos Hidráulicos, personal communication, August 1990.

66. Hermosillo 1988 : 95.
67. Sonora 1990*b* : 308–309.
68. Sonora 1990*a* : 97.
69. Martínez 1956 : 525, n. 43.
70. Freebairn 1963 : 1152.
71. Sonora 1984, II : 763–766.
72. Wellhausen 1976 : 128, 138.
73. Anon. 1983 : 356; Sonora 1990*b* : 8.
74. Harness and Barber 1964 : 1–2, 14; Hermosillo 1988 : 96.
75. Sonora 1984, II : 763–764.
76. Ibid.
77. Sonora 1990*a* : 93–94.
78. Smith 1947 : 1, 14; Cook 1966 : 1–2; Firch and Young 1968 : 6–7, 13; Emerson 1980*a* : 1, 5, 12.
79. Sonora 1990*a* : 93–94.
80. Emerson 1979 : 1–4, 9.
81. Emerson 1980*b* : 1–2.
82. Anon. 1988 : 7; Starkey 1980 : 209–210; Anon. 1990 : 12–13.
83. Dorronsoro 1964 : 104. According to Dorronsoro, in 1964, of the farms in the irrigated zones of the northwest coast of Mexico, including those of Sinaloa, Sonora, and the delta of the Colorado River in Baja California, 65.8 percent were totally mechanized, 32.5 percent, partially so, and only 1.6 percent nonmechanized. Today, the farms in the Costa de Hermosillo are nearly 100 percent mechanized.
84. Signoret Vera 1965 : 52.
85. Ibid.
86. Sonora 1990*b* : 155.
87. Dabdoub 1964 : 333–334.
88. Dozier 1963 : 562; Signoret Vera 1965 : 58; Henderson 1965 : 307; Spaulding 1974 : 380. In 1975 and 1976, following invasions of private holdings by landless peasants, more than forty thousand additional hectares of irrigated farms in the Yaqui Delta were given to *ejidatarios* in small individual parcels of ten hectares (Sanderson 1981 : 4–5).
89. Dozier 1963 : 562.
90. Rowan 1962 : 59; Dozier 1963 : 562; Henderson 1965 : 308.
91. Rowan 1962 : 59; Henderson 1965 : 306.
92. Sanderson 1981 : 149, 152.
93. Freebairn 1963 : 1158; Dozier 1963 : 564; Dabdoub 1964 : 374; Spaulding 1974 : 86.
94. Osorio Tafall 1943; Hubbs and Roden 1964 : 159, 185.
95. Edwards 1978 : 75.
96. Steinbeck and Ricketts 1941 : 246–247.
97. Only a small fishing camp in the 1920s, by 1950 Puerto Peñasco had grown to a bustling port of twenty-five hundred and fifty shrimp trawlers. Today it is a major fishing center of twenty-five thousand inhabitants (Ives 1950; Murray 1966 : 15–16; Mexico 1990 : Sonora, resultados preliminares). In 1920 Yávaros was a small port where the Standard Oil Company had facilities for storing fuel. The port was revived in the 1960s as a fishing center (Murray 1966 : 32).
98. McGuire 1986 : 137.
99. Young 1982 : 132–35; Treviño 1982 : 120.

100. Edwards 1978 : 146; Sonora 1990*b* : 100; Anon. 1971*b* : 148.

101. Edwards 1978 : 159; McGuire 1986 : 136.

102. McGuire 1986 : 123, 137; Mercado Sánchez 1961.

103. McGuire 1986 : 123.

104. Sonora 1990*b* : 11, 98–99.

105. Edwards 1978 : 175; Sonora 1990*b* : 100.

106. Hermosillo 1989 : 147–149.

107. Sonora 1990*b* : 11, 169.

108. Murray 1966 : 27.

109. Sonora 1990*a* : 228.

110. Rowan 1962 : 66; Dozier 1963 : 562; Anon. 1969 : 5; Cummings 1974 : 28. Today this agency is combined with the agriculture agency and called the Secretaría de Agricultura y Recursos Hidráulicos.

111. Sandoval Godoy 1988 : 133–203.

112. Sonora 1990*a* : 198–199.

113. Ibid. : 196.

114. Sonora 1990*b* : 123–124, 125–127.

115. South 1990 : 549–550.

116. Sonora 1990*b* : 45–47.

Bibliography

Original Documents

ARCHIVO FRANCISCANO, BIBLIOTECA NACIONAL DE MEXICO, MEXICO CITY

Legajo	*Short Title and Date*
12/200 bis exped. 7-27	Gregorio Alvarez Tuñón y Quirós sobre los asaltos de los Apaches, prov. de Sonora, 1704–1724.
34/738	Pedro de Corbalán sobre el estado de la provincia a su cargo. Alamos, 28 de enero de 1778.
35/764	Informe sobre San Antonio de la Huerta, 8 de octubre de 1788.
35/772	Plan que manifiesta las tierras de pan llevar y ranchos de la misión de San Miguel de Oposura y las misiones circunvecinas. 31 de agosto de 1790.
35/781	Enrique de Grimarest sobre el estado de los placeres de oro en la provincia de Sonora, 1791.
35/775	Enrique de Grimarest sobre el estado de los pueblos del Río Yaqui, 13 de agosto de 1790.
35/784	Enrique de Grimarest sobre el estado de las minas de Aigamé y Bayoreca, 16 de diciembre de 1791.
36/806	Sobre el estado deplorable de las misiones de Sonora, 8 de junio de 1797.
36/819	Informe sobre las provincias de Sonora y Ostímuri, 1804.
37/824	Informes del Real de San Ildefonso de Cieneguilla y sus alrededores, 1805–1809.
41/944	Enrique de Grimarest sobre los placeres de la Cieneguilla y otras minas de Sonora, 1791–1792.

ARCHIVO GENERAL DE INDIAS (AGI), SEVILLE, SPAIN

Ramo	*Legajo*	*Short Title and Date*
Escribanía	391-A	De la residencia que dió al gobernador Domingo de Terán de los Ríos, del tiempo que

« 162 »

		lo fué de los presidios de Sinaloa y Sonora, 1689.
	397-A	Libro de gobierno de títulos y mercedes que hizo el gobernador don Antonio de Oca Sarmiento en el Reyno de Nueva Vizcaya, 1666.
	400-A, cuaderno 1	Autos de visita general hecha en la prov. de Sonora por el capitán don Francisco Cuervo de Valdéz, visitador general, 1681.
	400-B, cuaderno 6	Informe de Francisco Cuervo de Valdéz y las minas de San Juan Bautista, Sonora, 1685–86.
	400-C, cuadernos 7, 9	Residencia a don Francisco Cuervo de Valdéz, alcalde mayor de Sonora, 1684.
Audiencia de Guadalajara	13	Juan López de Elisalde, mercader de Sonora, al presidente de la Audiencia acerca de alcabalas, 1674–1678.
	19	Testimonio de los autos hechos contra el general don Domingo de Terán de los Ríos, alcalde mayor de la provincia de Sinaloa, 1687.
	135	Sobre misiones y minas de Sonora y California; Informe de Juan Mateo Mange, 1735.
	137	Misiones de Sonora y California; minas de Sonora, 1751–1757.
	145	Expediente sobre si debía pagar el derecho de alcabala en las minas del Parral, 1673–1689.
	152	Expediente sobre los indios Tobosos y sus aliados, años de 1694 a 1698.
	154	Correspondencia de los vecinos del Valle de Sonora al gobernador de la Nueva Vizcaya, 1691.
	185	Expediente sobre las bolas de plata halladas en la Pimería Alta en la provincia de Sonora, 1737–1740.
	253	Informe sobre el estado de las Provincias Internas por Teodoro de Croix, 1780–1781.
	267	Sobre los placeres de oro del Real de Cieneguilla y las novedades de indios enemigos, Teodoro de Croix a José de Gálvez, 1779–1781.
	284	Estado de Sonora y extrato del padrón de la provincia, 1783
	296	Correspondencia de Nemesio Salcido, oficial de Cieneguilla a Miguel Cayetano Soler, sobre el estado de los placeres de Cieneguilla y alrededores, 1804–1807.
	416	Informes del visitador José de Gálvez sobre Sonora, 1771.

472	El Real Hacienda de Rosario y Alamos, 1772–1781.	« **163** »
505	Expedientes sobre la imposición y cobro de tributos in la Provincia de Sonora, 1776.	
512	Expediente sobre el Real de Cieneguilla y las Provincias Internas, 1771–1772.	
513	Expediente sobre el Real de Cieneguilla, 1772–1775.	
514	Expediente sobre el Real de Cieneguilla y los presidios de Sonora, 1773–1775.	
518	Noticia de los contratos que han celebrado varios comerciantes establecidos en la provincia de mi mando para abastecer las tropas. (Felipe de Neve to José de Gálvez, Oct. 20, 1783).	
519	Instrucción para el gobierno de la compañía de Fieles Opatas, pueblo de Bacoachi, y otros asuntos militares de la provincia de Sonora, 1784.	

Patronato   232, ramo 1   Papeles pertenecientes a la libertad y servicio personal de los indios en las provincias de Sonora y Sinaloa en el Nuevo Reino de Galicia, 1672–1679.

ARCHIVO GENERAL DE LA NACIÓN (AGN), MEXICO CITY

Ramo	Vol.	Expediente	Short Title and Date
Historia	16	—	Autos de visita de la provincia de Sonora, de don Tomás de Ugarte, 1673.
		—	Lic. José Rafael Rodríguez Gallardo al gobernador Diego Ortiz de Parrilla, Mátape, Marzo 15 de 1750.
	72	—	Memoria de Enrique de Grimarest, 1792.
	316	11	Sobre los reales de minas de Sonora, 1667.
		12	Memoria de los vecinos del real de San Miguel Arcángel, sobre la venta de trigo a los mineros de Sonora, 1673.
		14	Memoria contra los padres misioneros de Sonora por la venta de alimentos a los mineros, 1666.
Indios	40	29	Sobre el repartimiento de indios a los mineros de Sonora, por el padre Luis Mancusso, visitador de las misiones, 1715–1717.
Jesuitas	I	11	El alcalde mayor de San Juan Bautista al padre visitador, Luis Mancuso, 1715.
	II	10	Petición del vecindario de Sonora . . . para presidio a causa del alzamiento del Nuevo México . . . 1686.
		19	Instrucciones del Padre Visitador General de las misiones, 1717.

		92	Razón de la visita de este año de 1698, Colegio de Sinaloa, 1698.
Misiones	14	—	Informe del Obispo de Sonora, Antonio de los Reyes, 1772, 1784.
Provincias Internas	29	5	Autos sobre invasiones que hicieron los indios bárbaros en provincias de la Nueva Vizcaya, 1690–93.
		6	Informe que rinde al virrey el visitador de las Provincias Internas, don Joseph Bautista Rodríguez Gallardo, acerca del estado económico . . . de Sonora . . . 1750.
	30	5	Informes sobre el estado en que se halla la Provincia de Sonora, remetidos al gobernador Pardiñas Villar, por diversas personas, 1689.
		7	Autos sobre el estado en que se hallan los indios de Sinaloa y Río Yaqui, por el gobernador Pardiñas, 1689–1690.
		8	Informes remitidos al gobernador Pardiñas acerca del estado en que se halla la provincia de Sonora, 1689–1690.
	47	1	Memoria sobre San Antonio de la Huerta por don Buenaventura de Llenes, alcalde mayor de Sonora, 1767.
	48	3	Informes del comandante don Domingo Elizondo, de la visita de don José de Gálvez a las Provincias Internas; noticias de la situación particular de Sonora, 1767–1768.
	69	2	Copia íntegra del expediente para los establecimientos de presidios en la provincia de Nueva Vizcaya.
	70	1	Visita de don José de Gálvez a las Provincias Internas, 1768–1770.
	81	1	Correspondencia de don Domingo Elizondo con el virrey Marqués del Croix: estado de misiones, etc. de Sonora . . . , 1771–1772.
		18	Correspondencia de Pedro de Tueros, encargado de los reales de minas de Cieneguilla, provincia de Sonora, 1772–1773.
	82	1	Correspondencia diversa sobre Sonora, 1771–1772.
	86	1	Correspondencia de Bernardo de Urrea y otros comandantes de presidios de Sonora acerca del estado militar de la provincia, 1761–1765.
		2	El gobernador José de Tienda de Cuervo al Marqués de Cruillas sobre el estado de minas y placeres de oro en la provincia de Sonora, 1762.
	90	1	Correspondencia con el capitán Pedro de Tueros, comisionado en los placeres de Cieneguilla, 1774–1776.
	93	1	Correspondencia de Pedro Corvalán, intendente de Sonora, con el Virrey; informes acerca de los célebres placeres de Cieneguilla, 1771.
	152	1	Diversos informes de los religiosos que tienen a su cargo

		las misiones de las Provincias Internas que dejaron los « 165 » Jesuitas cinco años antes, 1772.
226	2	Correspondencia de don Pedro Corvalán, sub-intendente de la Expedición de Sonora, sobre sus ministros a las tropas, 1767–1770.
232	1	Correspondencia de don Pedro Corvalán, intendente de Sonora, sobre el descubrimiento de los placeres de Cieneguilla, los indios Seris, etc., 1772–1773.
245A	13	Memoria del padre Ignacio Lizossoain sobre rebeliones indígenas; sobre San Antonio de la Huerta, etc., 1763.
	23	Sobre los placeres del real de Cieneguilla, 1771–1772.
246	21	Don Pedro de Corbalán, sobre la Expedición de Sonora, 1776–1777.
	23	Capitán Pedro de Tueros rinde informes sobre el real de minas de Palo Encebado, 1776–1777.
247	16	Sobre las minas de San Marcial, etc., 1773–1775.
	18	Padrón de vecinos del Real de Cieneguilla, 1773–1775.
259	7	Informe de don Pedro Garrido y Durán, intendente y gobernador de Sonora y Sinaloa, sobre el estado de la administración de aquellas provincias, 1790.

ARCHIVO HISTÓRICO DE HACIENDA (AHH), MEXICO CITY

(Ramo de Temporalidades)

Legajo	*Expediente*	*Short Title and Date*
17	42	Breve resumen de los desastres . . . acaecidos en la provincia de Sonora hostilizado de los Apaches, Seris y Pimas alzados, 1755–1760.
278	3	Sobre la venta de cosechas de la misión de Mátape a los mineros de Sonora, 1681, 1695.
	11	Sobre los insultos que hicieron los indios enemigos contra los reales de minas de Nacazori, Bacanuchi, Basachuca, etc. . . . 1715.
	20	Visita de las misiones en la provincia de Sonora, 1729, 1749.
	24	Manifestaciones del general don Pedro Francisco Sartillán, alcalde que fué de la provincia de Sonora y otros testigos del Real de San Miguel Arcángel, 1673.
	36	Manifestaciones del padre Daniel Angelo Marras, rector del Colegio de Mátape, sobre la venta de cosechas a los mineros de Sonora, 1671.
	41	Memoria de Joseph Martín Genovés, visitador de misiones de Sonora, 1722.
	50	Sobre el comercio y rutas de transporte de la provincia de Sonora al Real del Parral. Correspondencia de varios padres, 1680–89.
279	28	Razón de la hacienda de San Joseph de Tecoripa [hacienda de

		sacar plata de los padres de Mátape], n.d. [ca. 1690].
	29	Inventario de los bienes de partido de Mátape, 1690.
282	11	Memoria sobre la venta de cosechas a los mineros de Sonora, por los padres misioneros, 1690, 1701; relación sobre los pimas.
325	8	Carlos de Rojas a padre provincial Francisco Cavallos, sobre el estado de Sonora. Arizpe, 1765.
	24	Pedro Daniel Januski al alcalde mayor de la provincia de Sonora, Francisco Pacheco, sobre los hacendados españoles en Sonora, 1713.
	33	Padre Daniel Januski sobre las invasiones de los hacendados dentro de las tierras de los indios, 1713.
	69	Sobre las ventas de productos agrícolas de las misiones a los mineros de Sonora, 1673.
	71	Relación que presentó en 21 de agosto de 1673 de Gregorio López de Dicastillo.
	87	Mandamientos dados por los alcaldes mayores y tenientes . . . mandando a los alcaldes y oficiales de los pueblos que den indios a los mineros y a otras personas, 1713.
377	73	Bienes de la misión de Sahuaripa, 1735.
974	1	Diego Ortiz Parilla a don Juan Francisco de Guemes y Horcasitas, 1750.
1126	1	Requerimiento que el padre Leonardo Jatino, visitador de las misiones de la Compañía de Jesús en la provincia de Sonora, hace a don Pedro de Oliva, teniente del gobernador capitán don Pedro de Perea en el puesto de Nombre de Dios, 1640.
2016	2	Memoria del Padre Felipe Segesser, ministro de la misión de Tecoripa.

ARCHIVO MUNICIPAL DEL PARRAL (AMP), PARRAL, CHIHUAHUA

Legajo and Year	*Expediente*	*Short Title*
1649 Administración	13	Autos hechos en razón de la entrada de Curcuribazca y Buchibacuachi en la prov. de Sonora, hechos por el capitán Simón Lazo de la Vega.
1658 Administración	4	Expediente que contiene algunas cédulas y certificaciones de la plata que ha salido de Sonora.
1663 Administración	104	Legajo de cédulas de mineros de la provincia de Sonora y certificaciones de la justicia mayor, de la plata que traen a este Real de Parral . . .
1664 Administración	13	Legajo de manifestaciones de plata de Sonora.
1666 Administración	—	Legajo de manifestaciones de plata de Sonora.
1669 Administración	4	Legajo de manifestaciones de plata de Sonora.
1673 Administración	118	Información a solicitud de Nicholás de Mora, para comprobar que tiene hacienda de sacar plata en la Prov. de Sonora.

1677 Administración	88	Residencia que se tomó al alcalde mayor don Alonzo Razón al tiempo que ministró justicia en San Juan Bautista.	« 167 »

1677 Administración 133 Manifestaciones de plata de Sonora.

1678 Administración 104 Autos de visita que practicó a las provincias de Sonora y Sinaloa el general don Francisco Fuentes y Sierra.

1682 Administración 7 Autos de visita practicados por el general don Francisco Cuevo de Valdés en la prov. de Sonora.

1685 Administración 102 Autos de visita practicados por el general don Gabriel de Isturiz . . . en la prov. de Sonora.

1688 Administración 139 Residencia que se tomó a don Antonio Barba Figueroa del tiempo que administró justicia como alcalde mayor de la prov. de Sonora.

1707 Administración 12 Diligencias practicadas contra el alcalde mayor de San Juan Bautista de Sonora por desobediencia a unos mandamientos.

1715 Administración 136 Despacho del Sr. Virrey de la Nueva España, Duque de Linares, por las justicias de la prov. de Sonora.

1718 Administración 2 Visita de la prov. de Sonora por el gobernador del Reyno [Nuevo Vizcaya].

1723 Administración 103 Autos de visita del Valle de San José de Opodepe, por el capitán don Juan Antonio Fernández de la Cabada . . .

107 Autos de visita al real presidio de Conchos, el real de San Juan Bautista y al real de San Antonio de Motepore, por el general don Martín de Alday.

1720 Civil 108 Demanda del general Andrés Baca Guajardo, vecino de Durango, contra Antonio de Orante, vecino de Ostímuri, provincia del territorio de Sonora.

1728 Civil 123 Juicio seguido por el alférez Sebastián López de Guzmán en el real de minas de Potreros contra Patricio Nájera por entrega de bienes.

1688 Minas 139 Manifestaciones de plata, prov. de Sonora, 1677–1688.

1675 Terrenos y Sitios 20 Título de un sitio de tierra en la prov. de Sonora en favor del alférez Juan López de Pérez.

ARCHIVO HISTÓRICO DEL GOBIERNO DEL ESTADO DE SONORA (AHG-SONORA), HERMOSILLO

Ramo minería, expediente Distrito de Altar [late eighteenth and early nineteenth centuries].

« **168** » BANCROFT LIBRARY, BERKELEY, CALIFORNIA

Pinart Papers, Documents relating to Northern Mexico, ser. 1, items 23, 24.

BENSON LATIN AMERICAN COLLECTION,
UNIVERSITY OF TEXAS, AUSTIN

W. B. Stevens Collection no. 68: Noticia de las misiones que administran los padres de la Compañía de Jesús en esta Nueva España, año de 1765.

BIBLIOTECA NACIONAL DE MADRID (BNM-MADRID),
MANUSCRIPT DIVISION

MS 19266   Descripción sucinta de las provincias de Culiacán, Sinaloa y Sonora, 1772.

BIBLIOTECA NACIONAL DE MÉXICO, MANUSCRIPT DIVISION
(BHM-MS), MEXICO CITY

MS 489   Apología en defensa de los padres de la Compañía de Jesús de la Provincia de Sonora por Joseph de Pallares, 8 de enero de 1705.

BIBLIOTECA PÚBLICA DEL ESTADO, GUADALAJARA
(MANUSCRIPT DIVISION)

Ramo Civil (1700–1719).

Published and Unpublished Materials

Acosta, Roberto
  1946   "Ciudad de Alamos." *Memorias de la Academia Mexicana de Historia* 5:38–67.
Alegre, Francisco Javier
  1841–1842   *Historia de la Compañía de Jesús en Nueva-España al tiempo de su expulsión.* 3 vols. Mexico City: J. M. Lara.
Almada, Francisco R.
  1952   *Diccionario de historia, geografía y biografía sonorenses.* Chihuahua: Ruiz Sandoval.
Amsden, Monroe
  1928   *Archaeological Reconnaissance in Sonora.* Southwestern Museum Papers, no. 1. Los Angeles.
Anderson, T. H., et al.
  1970   "Reconnaissance Survey of Precambrian Rocks, Northwestern Sonora." *Abstract of Papers*, Geological Society of America, Annual Meeting (Milwaukee, 1970), p. 484.
Anonymous
  1943   "Avance de las obras de la Comisión Nacional de Irrigación durante el año de 1942." *Irrigación en México* 24(1):12–56.
  1961   "Areas de tierras de riego ensalitradas de México." *Ingeniería Hidráulica en México* 15(1):71.
  1967   "Veinte años de labores de la Secretaría de Recursos Hidráulicos" [1947–1967]. *Ingeniería Hidráulica en México* 21(1):17–58.

1969	"Regional Analysis. Northwestern: Baja California, Sonora and Sinaloa." *Review of the Economic Situation of Mexico* 45(518) : 3–7.
1970	"Notas sobre Sonora: Caborca—Francisco Xavier Saeta, S.J.—La matanza de El Tupo." *Memorias de la Academia Mexicana de Historia* 29: 342–349.
1971*a*	"Dirección de Agrología. Informe de actividades de mayo de 1967 a septiembre de 1970." *Ingeniería Hidráulica en México* 25(1) : 71–77.
1971*b*	"Sonora, a Rich State on the Move." *Review of the Economic Situation of Mexico* 47(4) : 142–150.
1983	"Notes on Mexican Wheat." *Review of the Economic Situation of Mexico* 59(11) : 354–59.
1988	"U.S. Becomes Major Wheat Supplier to Mexican Market." *Foreign Agriculture* 26(8) : 7
1990	*Mexico, Country Report, Economist Intelligence Unit.* No. 1 : 12–18; no. 4 : 16–17.

Bahre, Conrad J.
1967	The Reduction of Seri Indian Ranges and Residence in the State of Sonora, Mexico, 1536–Present. M.A. thesis, University of Arizona.

Bahre, Conrad J., and Bradbury, David E.
1980	"The Manufacture of Mescal in Sonora, Mexico." *Economic Botany* 34: 391–400.

Bancroft, Hubert H. (ed.)
1884	*History of the North Mexican States and Texas*, Vol. I, 1531–1800. San Francisco: A. L. Bancroft & Co.
1889*a*	*History of the North Mexican States and Texas*, Vol. II, 1801–1889. San Francisco: A.L. Bancroft & Co.
1889*b*	*History of Arizona and New Mexico.* San Francisco: The History Co.

Bannon, John F.
1955	*The Mission Frontier in Sonora, 1620–1687.* New York: U.S. Catholic Historical Society.

Bartlett, John R.
1852	*Report of the Secretary of the Interior . . . Certain Papers in Relation to the Mexican Boundary Commission.* Senate Doc. no. 6, Special Session, 33rd Congress, pp. 94–103.

Benassini, Aurelio
1942	"Potencialidad agrícola de la región costera de los estados de Sonora y Sinaloa." *Irrigación en México* 23(5) : 89–103.

Bernstein, Marvin D.
1964	*The Mexican Mining Industry, 1890–1950.* Albany: State University of New York.

Blackmore, F. W.
1891	*Spanish Institutions of the Southwest.* Baltimore: Johns Hopkins University Press.

Blake, Dean
1935	"Mexican West-Coast Cyclones." *Monthly Weather Review* 63 : 344–348.

« 170 »   Bloomer, George W.
    1909   "Mining in the Alamos and Arteaga Districts." *Engineering and Mining Journal* 87 : 699–700.

Bolton, Herbert E.
    1919   *Kino's Historical Memoir of Pimeria Alta.* 2 vols. Cleveland: Arthur H. Clark Co.
    1936   *Rim of Christendom: A Biography of Eusebio Francisco Kino, Pacific Coast Pioneer.* New York: Macmillan Co.

Bolton, Herbert E., and Marshall, Thomas M.
    1920   *The Colonization of North America 1492–1783.* New York: Macmillan Co.

Bond, J. B.
    1935   "Proyecto del Río Yaqui. Informe rendido por el Ing. J. B. Bond en el año de 1929." *Irrigación en México* 10(4) : 203–230.

Bonillas, Y. S.
    1911   "Algunos datos geológicos sobre el mineral de La Campana, Distrito del Altar, Sonora." *Boletín de la Sociedad Geológica de México* 7 : 155–168.

Borthwick, J. D.
    1857   *Three Years in California* (1851–54). Edinburgh & London: W. Blackwood & Sons.

Braniff, Beatriz
    1978   "Preliminary Interpretations Regarding the Role of the San Miguel River, Sonora, Mexico." In C. Riley, and B. Hedrick (eds.), *Across the Chichimec Sea*, pp. 67–82.
    1984   "Proyecto Río San Miguel, Sonora." *Boletín del Consejo de Arqueología, 1984*, (edited by Joaquín García Barcena), pp. 9–21. Mexico City: Instituto Nacional de Antropología e Historia.

Brown, David E. (ed.)
    1982   "Biotic Communities of the American Southwest—United States and Mexico." *Desert Plants* 4 (1–4).

Brown, George W.
    1968–1974 *Desert Biology.* 2 vols. New York: Academic Press.

Burrus, Ernest J.
    1971   *Kino and Manje, Explorers of Sonora and Arizona.* Rome & St. Louis: Jesuit Historical Institute.

Busch, C. D., et al.
    1966   "Utilization of Water Resources in a Coastal Groundwater Basin. Part I, Evaluation of Irrigation Efficiency." *Journal of Soil and Water Conservation* 21(5) : 163–166.

Cajeme [Sonora]
    1989   *Agenda estadística municipal, 1988.* Ciudad Obregón: H. Ayuntamiento de Cajeme, Sonora.

Calderón, Francisco R.
    1965   "Los ferrocarriles." In D. Cosío Villegas (ed.), *Historia moderna de México*, vol. 5, pt. 1, pp. 483–634.

Carta topográfica, 1 : 50,000, Estados Unidos Mexicanos. (sheet G12B44, Villa Juárez [Sonora], 1975.

Castetter, Edward F., and Willis H. Bell
1942 *Pima and Papago Indian Agriculture*. Albuquerque: University of New Mexico Press.

Colección Mateu (Barcelona). Typescript copies of original documents on Sonora, in Special Collections, University of Arizona Library, Tucson.

Cook, A. Clinton
1966 *Survey of Mexican Vegetable and Melon Production*. Washington, D.C.: U.S. Dept. of Agriculture, Foreign Agriculture Service, FAS-M:178.

Cosío Villegas, Daniel (ed.)
1955 *Historia moderna de México*. 9 vols. Mexico City: Editorial Hermes.

Crumrine, N. Ross
1977 *The Mayo Indians of Sonora: A People Who Refuse to Die*. Tucson: University of Arizona Press.

Cserna, Zoltan de
1970 "The Precambrian of Mexico." In *The Geologic Systems*, vol. 4, pp. 253–270. New York & London: Interscience–John Wiley & Sons.

Cummings, Ronald G.
1974 *Interbasin Water Transfers. A Case Study in Mexico*. Baltimore: Johns Hopkins University Press.

Dabdoub, Claudio
1964 *Historia de El Valle del Yaqui*. Mexico City : Librería de Manuel Porrúa.

Damon, Paul E., et al.
1962 "Extension of the Older Precambrian of the Southwest into Sonora, Mexico." Geological Society of America, *Special Papers*, no. 68, pp. 158–159.

Darby, H. Clifford
1953 "On the Relation of Geography and History." *Transactions and Papers of the Institute of British Geographers* 19:1–12.

Diccionario Porrúa
1964 *Historia, biografía y geografía de México*. Mexico City: Editorial Porrúa.

Dice, L. B.
1939 "The Sonoran Biotic Province," *Ecology* 20:118–129.

Di Peso, C. C., and Mason, D. S.
1965 "The Seri Indians in 1692: As Described by Adamo Gilg, S.J." *Arizona and the West* 7:33–56.

Documentos para la historia de México. Ser. 3 and 4, 1853–1857. 3 vols. Mexico City: Imprenta de J. R. Navarro.

Dodge, Richard N.
1973 Morphology and Function of the Road Network of Eastern Sonora. M.A. thesis, University of Arizona.

D'Olwer, Luis N.
1965 "Las inversiones extranjeras." In D. Cosío Villegas (ed.), *Historia moderna de México*, vol. 5, pt. 2, pp. 973–1185.

Donohue, J. A.
1969 *After Kino: Jesuit Missions in Northwestern New Spain, 1711–1767*. Rome & St. Louis: Jesuit Historical Institute.

« 172 »   Doolittle, William E.

1984*a*   "Settlements and the Development of 'Statelets' in Sonora, Mexico."
Journal of Field Archaeology 11:13–24.

1984*b*   "Cabeza de Vaca's Land of Maize: An Assessment of Its Agriculture."
Journal of Historical Geography 10:246–262.

1988   *Pre-Hispanic Occupance in the Valley of Sonora, Mexico*. Anthropological
Papers, University of Arizona, no. 48.

Dorronsoro, José María

1964   "La mecanización de la agricultura en los distritos de riego en México."
Ingeniería Hidráulica en México 18(1):102–111.

Dozier, Craig L.

1963   "Mexico's Transformed Northwest: The Yaqui, Mayo, and Fuerte Examples." *Geographical Review* 53:548–571.

Dunbar, Gary S.

1974   "Geographical Personality." *Geoscience and Man* 5:25–33.

Edwards, R. R. C.

1978   "Ecology of a Coastal Lagoon Complex in Mexico." *Estuarine and
Coastal Marine Science* 6:75–92.

Emerson, L. P. Bill, Jr.

1979   *Mexico's Grape Industry: Table Grapes, Raisins and Wine*. Washington,
D.C.: U.S. Dept. of Agriculture, Foreign Agriculture Ser., M292.

1980*a*   *Mexico's Vegetable Production for Export*. Washington, D.C.: U.S. Dept.
of Agriculture, Foreign Agriculture Ser. M279.

1980*b*   *Mexico's Expanding Olive Industry*. Washington, D.C.: U.S. Dept. of Agriculture, Foreign Agriculture Ser. M294.

Estadísticas históricas de México

1986   2 vols. Mexico City: Instituto Nacional de Estadística, Geografía e
Información.

Evans, Emyr Estyn

1973   *The Personality of Ireland: Habitat and History*. Cambridge: University
Press. (Enlarged edition, 1981. Belfast: Blackstaff Press.)

Fabila, Alfonso

1940   *Las tribus Yaquis de Sonora: su cultura y anhelada autodeterminación*. Mexico City: Departamento de Asuntos Indígenas.

Fabry, José Antonio

1743   *Compendiosa demostración de los crecidos adelantamientos que pudiera lograr
la Real Hacienda de su Magestad, mediante la rebaja en el precio de
azogue* . . . Mexico City: Por la viuda de J. B. Hogal.

Fehrenbach, T. R.

1974   *Comanches: The Destruction of a People*. New York: Alfred A. Knopf.

Felger, Richard S.

1980   "Vegetation and Flora of the Gran Desierto," *Desert Plants* 2:87–114.

Felger Richard S., and M. B. Moser

1970   "Seri Use of *Agave* (Century Plant)." *The Kiva* 35(4):159–167.

1976   "Seri Indian Food Plants: Subsistence without Agriculture." *Ecology of
Food and Nutrition* 5(1):13–17.

Firch, Robert S., and Robert A. Young   « 173 »

1968   *An Economic Study of the Winter Vegetable Export Industry of Northwest
        Mexico.* Agricultural Experiment Station, Technical Bulletin no. 179.
        Tucson: University of Arizona.

Fontana, Bernard L.

1974   "Man in Arid Lands: The Piman Indians of the Sonoran Desert." In
        G. W. Brown (ed.), *Desert Biology*, Vol. 2, pp. 489–528.

Forbes, Jack D.

1957   "Historical Survey of the Indians of Sonora, 1821–1910." *Ethnohistory*
        4:335–368.

1959   "Unknown Athapaskans: The Identification of the Jano, Jacome, Ju-
        mano, Manso, Suma, and Other Indian Tribes of the Southwest."
        Ethnohistory 6:97–159.

1960   *Apache, Navaho and Spaniard.* Norman: University of Oklahoma Press.

Freebairn, Donald K.

1963   "Relative Production Efficiency between Tenure Classes in the Yaqui
        Valley, Sonora, Mexico." *Journal of Farm Economics* 45(4):1150–1160.

Galaviz de Capdevielle, María Elena

1968   "Los apaches en los siglos XVII y XVIII." *Memorias de la Academia Mex-
        icana de Historia* 27:5–14,

Gentry, Howard S.

1942   *Rio Mayo Plants: A Study of the Flora and Vegetation of the Valley of the
        Rio Mayo, Sonora.* No. 527. Washington, D.C.: Carnegie Institution.

Gerhard, Peter

1972   *A Guide to the Historical Geography of New Spain.* Cambridge: At the
        University Press.

1982   *The North Frontier of New Spain.* Princeton: Princeton University Press.

Goldman, E. A., and R. T. Moore

1945   "The Biotic Provinces of Mexico." *Journal of Mammalogy* 26(4):
        347–360.

Griffen, William B.

1985   "Problems in the Study of Apache and Other Indians in Chihuahua and
        Southern New Mexico during the Spanish and Mexican Periods." *The
        Kiva* 50:139–151.

1988a   *Apaches at War and Peace. The Janos Presidio, 1750–1858.* Albuquerque:
        University of New Mexico Press.

1988b   *Utmost Good Faith: Patterns of Apache-Mexican Hostilities in Northern Chi-
        huahua Border Warfare, 1821–1848.* Albuquerque: University of New
        Mexico Press.

Guinn, James M.

1909–1910 "The Sonoran Migration." *Annual Publication of the Historical Society of
        Southern California* 8:31–36.

Gulhati, N. D.

1958   "Worldwide View of Irrigation Developments." *Journal of Irrigation and
        Drainage Division* (Proceedings of the American Society of Civil Engi-
        neers) 84(3):1–14.

« **174** »   Hafer, Claude
  1912   "The Mines of the Sonora Valley, Mexico." *Mining and Engineering World* (April 17):903–904.
Hamilton, Patrick
  1884   *The Resources of Arizona.* 3rd. ed. San Francisco: A. L. Bancroft & Co.
Harlem, Arthur D.
  1964   "Cerro de Trincheras: Its History and Purpose." *Tlalocan* 4:339–350.
Harness, Vernon L., and Charles H. Barber
  1964   *Cotton in Mexico.* Washington, D.C.: U.S. Dept. of Agriculture, Foreign Agriculture Service, FAS-M 163.
Haskell, J. Loring
  1987   *Southern Athapaskan Migration,* A.D. *200–1750.* Tsaile, Ariz.: Navajo Community College Press.
Hastings, James R., and R. R. Humphrey (eds.)
  1969   *Climatological Data and Statistics for Sonora and Northern Sinaloa.* Technical Reports on Arid Regions, no. 19. Tucson: University of Arizona.
Hastings, James R., and Raymond M. Turner
  1965   *The Changing Mile.* Tucson: University of Arizona Press.
Heikes, V. V., and C. G. York
  1913   "Dry Placers in Arizona, Nevada, New Mexico and California." *Mineral Resources in the United States* [1912], pt. 1:254–263. Washington, D.C.: United States Geological Survey.
Henderson, David A.
  1965   "Arid Lands under Agrarian Reform in Northwest Mexico." *Economic Geography* 41:300–312.
Hermosillo (Sonora)
  1989   *Agenda estadística de 1988.* Hermosillo: H. Ayuntamiento de Hermosillo.
Hernández, Salvador V.
  1971   "Presidios en Sonora el año de 1764." *Boletín del Archivo General de la Nación* (ser. 2) 12:29–58.
Hu-DeHart, Evelyn
  1981   *Missionaries, Miners and Indians: Spanish Contact with the Yaqui Nation of Northwestern New Spain, 1533–1820.* Tucson: University of Arizona Press.
  1984   *Yaqui Resistence and Survival: The Struggle for Land and Autonomy, 1821–1910.* Madison: University of Wisconsin Press.
Hubbs, Carl L., and Gunnar I. Roden
  1964   "Oceanography and Marine Life along the Pacific Coast of Middle America." *Handbook of Middle American Indians*, vol. 1, Natural Environment and Early Cultures (R. C. West, vol. ed.), pp. 143–186. Austin: University of Texas Press.
Humboldt, Alexander von
  1811   *Political Essay on the Kingdom of New Spain.* 2 vols. (English translation by John Black). New York: I. Riley.
Ives, Ronald L.
  1950   "Puerto Peñasco, Sonora." *Journal of Geography* 49:349–361.

Jardines Moreno, José Luis
  1976   "Los distritos de riego por bombeo del centro y norte de México." *Recursos Hidráulicos* 5(1) : 8–25.
Jiménez Villalobos, Angel
  1965   "Condiciones de las aguas subterráneas en el distrito de riego número 51, Costa de Hermosillo, Sonora." *Ingeniería Hidráulica en México* 19(3) : 65–80.
Johnson, Alfred
  1963   "The Trincheras Culture of Northern Sonora." *American Antiquity* 29 : 174–186.
Johnson, Jean B.
  1950   *The Opata: An Inland Tribe of Sonora.* Albuquerque: University of New Mexico Publications in Anthropology, no. 6.
Kalstrom, George W.
  1952   "El Cordonazo—The Lash of St. Francis." *Weatherwise* 5 : 99–103.
King, Robert E.
  1939   "Geological Reconnaissance in Northern Sierra Madre Occidental of Mexico." *Bulletin of the Geological Society of America* 50 : 1625–1722.
Lavisse, Ernest (ed.)
  1900–1911 *Histoire de la France illustrée depuis les origines jusqu'a la revolution.* 9 vols. Paris: Hachette. Reprint 1941.
Leopold, A. Starker
  1952   *Wildlife of Mexico: The Game Birds and Mammals.* Berkeley & Los Angeles: University of California Press.
Livingston, Donald E., and Paul E. Damon
  1968   "The Ages of Stratified Precambrian Rock Sequences in Central Arizona and Northern Sonora." *Canadian Journal of Earth Sciences* 5 : 763–772.
Lockwood, Frank C.
  1934   *With Padre Kino on the Trail.* Tucson: University of Arizona Social Science Bulletin, vol. 5, no. 2.
  1938   *The Apache Indians.* New York: Macmillan Co.
Logan, Walter S.
  1894   *Yaqui, the Land of Sunshine and Health: What I Saw in Mexico.* New York: Albert B. King.
Loma, José Luis de la
  1963   "Características en los distritos de riego en México." *Ingeniería Hidráulica en México* 18(1) : 17–26.
Love, Frank
  1974   *Mining Camps and Ghost Towns: A History of Mining in Arizona and California along the Lower Colorado.* Los Angeles: Westernlore Press.
Lumholtz, Karl S.
  1912   *New Trails in Mexico; an Account of One Year's Exploration in Northwestern Sonora, Mexico, and Southwestern Arizona, 1909–1910.* New York: Scribner.
Lumholtz, Karl S., and I. N. Dracopoli
  1912   "The Sonora Desert." *Geographical Journal* 40 : 503–518.

« 176 »   McGee, William J.
     1897   "Sheetflood Erosion." *Bulletin of the Geological Society of America*
                   8:87–112.
          McGuire, Thomas R.
     1986   *Politics and Ethnicity on the Rio Yaqui: Potam Revisited*. Tucson: Univer-
                   sity of Arizona Press.
          Mails, Thomas E.
     1974   *The People Called Apache*. Englewood Cliffs, N.J.: Prentice-Hall.
          Mange, Juan Matheo
     1926   [1720] *Luz de tierra incógnita en la América septentrional y diario de las
                   exploraciones en Sonora*. Publicaciones del Archivo General de la Nación,
                   no. 10. Mexico City.
          Martínez, Pablo L.
     1956   *Historia de Baja California*. Mexico City: Libros Mexicanos.
          Matlock, W. G., et al.
     1966   "Utilization of Water Resources in a Coastal Groundwater Basin." Pt. 2,
                   "Groundwater Supply and Incipient Saltwater Intrusion." *Journal of Soil
                   and Water Conservation* 21(5):166–69.
          Mendizábal, Miguel Othón de
     1930   *Evolución del noroeste de México*. Mexico City: Imprenta Mundial.
          Mercado Sánchez, Pedro
     1961   Corrección y modernización del sistema de captura del camarón en
                   aguas interiores del noroeste de México." *Acta Zoológica Mexicana*.
                   4(5):1–11.
          Merriam, Willis B.
     1957   "Irrigation Progress in the Mexican Northwest." *Journal of Geography*
                   56:429–433.
          Merrill, F. J. H.
     1905   "Geology of Sonora." *Engineering and Mining Journal* 80:976.
     1906   "The Mines of Planchas de Plata." *Engineering and Mining Journal*
                   82:1111-1112.
     1908a   "Surface Enrichment in Sonora." *Mining and Scientific Press* (June 13):
                   802–803.
     1908b   "Dry Placers of Northern Sonora." *Mining and Scientific Press* (Septem-
                   ber 12):360–361.
          Mexico
     1900   *Censo general de la Repùbica Mexicana*. Mexico City: Dirección General
                   de Estadística.
     1948   "Informe de labores de la Secretaría de Recursos Hidráulicos, 1 sep-
                   tiembre 1947 a 31 agosto de 1948." *Ingeniería Hidráulica en México*
                   2(3):25–185.
     1990   "Censo de población y vivienda, Sonora (resultados preliminares)." Her-
                   mosillo: Instituto Nacional de Estadística, Geografía e Información.
          Mowry, Sylvester
     1864   *Arizona and Sonora; the Geography, History, and Resources of the Silver
                   Region of North America*. 3rd. ed. New York: Harper & Brothers.
          Murray, Spencer
     1966   *Boating and Fishing Guide to the West Coast of Mexico*. Studio City, Calif.:
                   Vaquero Books.

Nabhan, Gary P., and Richard S. Felger   « 177 »
1978   "Teparies in Southwestern North America: A Bibliographical and Ethno-
historical Study of *Phaseolus acutifolius.*" *Economic Botany* 32:2–19.
Nabhan, Gary P., et al.
1979   "Legumes in the Papago-Pima Indian Diet and Ecological Niche." *The
Kiva* 44(2–3):173–190.
Navarro García, Luis
1964   *Don José de Gálvez y la comandancia general de las Provincias Internas.*
No. 148. Seville: Escuela de Estudios Hispano-Americanos.
1965a   *Las provincias internas en el siglo XIX.* No. 162. Seville: Escuela de Estudios Hispano-Americanos.
1965b   "La sublevación Yaqui en 1740." *Anuario de Estudios Americanos* 22:
373–531.
1967   *Sonora y Sinaloa en el siglo XVII.* No. 176. Seville: Escuela de Estudios
Hispano-Americanos.
Naylor, Thomas H., and Charles Polzer
1986   *The Presidio and Militia on the Northern Frontier of New Spain.* Tucson:
University of Arizona Press.
[Nentvig, Juan Bautista]
1957–1958   "Descripción geográfica, natural y curiosa de la Provincia de Sonora, por
un amigo del servicio de Dios y del Rey, nuestro Señor. Año de 1764."
Boletín del Archivo General de la Nación [Mexico] 28:515–530, 659–705;
29:37–68.
1980   *Rudo Ensayo: A Description of Sonora and Arizona in 1764.* Translated and
annotated by Alberto Pradeau and Robert Rasmussen. Tucson: University of Arizona Press.
Ocaranza, Fernando de
1937–1939   *Crónicas y relaciones del occidente de México.* 2 vols. Mexico City: Antigua
Librería Robredo de J. Porrùa e Hijos.
Orive Alba, Adolfo
1945   "La política de irrigación." *Irrigación en México* 26(1):7–41.
Orozco y Berra, Manuel
1857   "Informe sobre la acuñación de las casas de moneda de la república." In
G. Manuel Siliceo (ed.), *Memoria de la Secretaría de Fomento, Colonización, Industria y Comercio de la República Mexicana.*
Osorio Tafall, B. F.
1943   "El Mar de Cortés y la productividad fitoplanctónica de sus aguas." *An-
ales de la Escuela Nacional de Ciencias Biológicas* 3:73–118.
Pailes, Richard A.
1978   "The Rio Sonora Culture in Prehistoric Trade Systems." In C. Riley,
and B. Hedrick (eds.), *Across the Chichimec Sea,* pp. 134–43.
1980   "The Upper Rio Sonora Valley in Prehistoric Trade." *Transactions, Illi-
nois State Academy of Science* 72(4):20–39.
Park, Joseph H.
1962   "Spanish Indian Policy in Northern Mexico, 1765–1810." *Arizona and the
West* 4:325–344.
Pearce, W. D.
1911   "Mining in the Alamos District." *Engineering and Mining Journal*
92:687–89.

« 178 »   Pennington, Campbell W.

1980   *The Pima Bajo of Central Sonora, Mexico*. 2 vols. Vol. 1: *The Material Culture*. Salt Lake City: University of Utah Press.

Pérez de Ribas, Andrés

1645   *Historia de los triumphos de nuestra santa fe entre gentes las más bárbaras, y fieras del Nueuo orbe* . . . Madrid: A de Paredes.

1944   *Triunfos de nuestra santa fe entre gente las más bárbaras y fieras del Nuevo orbe* . . . 3 vols. Mexico City: Editorial "Layac."

1968   *My Life among the Savage Nations of New Spain*. Edited in condensed form by Tomás Antonio Robertson. Los Angeles: Ward Ritchie Press.

1985   *Páginas para la historia de Sonora: Pérez de Rivas, triunfos de nuestra Santa Fe*. 2 vols. Hermosillo: Gobierno del Estado de Sonora.

Pfefferkorn, Ignaz

1949   *Sonora, a Description of the Province*. Translated and annotated by Theodore Treutlein. Albuquerque: University of New Mexico Press.

Pfeifer, Gottfried

1939   "Sinaloa und Sonora. Beiträge zur Landeskunde und Kulturgeographie des nordwestlichen Mexiko." *Mitteilungen der Geographischen Gesellshaft in Hamburg* 46 : 289–460.

Polzer, Charles W.

1972   "The Evolution of the Jesuit Mission System in Northwestern New Spain, 1600–1767." Ph.D. dissertation, University of Arizona.

1976   *Rules and Precepts of the Jesuit Missions of Northwestern New Spain*. Tucson: University of Arizona Press.

Pradeau, Alberto F., and Ernest J. Burrus (comps.)

N.d.   Historia de las misiones sonorenses y sus pueblos de visitas. Typescript, Arizona Collection, University of Arizona Library, Tucson.

Radding, Cynthia

1990   "Ethnicity and the Emerging Peasant Class of Northwestern New Spain, 1760–1840." Ph.D. dissertation, University of California, San Diego.

Ramírez, José Carlos (ed.)

1988   *La nueva industrialización en Sonora: el caso de los sectores de alta tecnología*. Hermosillo: Colegio de Sonora.

Ramírez, Santiago

1884   *Noticias históricas de la riqueza minera de México*. Mexico City: Secretaría de Fomento.

Recopilación de leyes de los reynos de las Indias.

1943   Mandadas imprimir publicar por la Magestad católica del rey don Carlos II, nuestro señor . . . 3 vols. Madrid: La viuda de J. Ibarra, impresor, 1791. Reprint, Madrid: Gráficas Ultra.

Reff, Daniel T.

1991   *Disease, Depopulation, and Culture Change in Northwestern New Spain, 1518–1764*. Salt Lake City: University of Utah Press.

Reyes, Antonio de los (OFM)

1938   "Memorial sobre las misiones de Sonora [julio 6 de 1772]." *Boletín del Archivo General de la Nación* [Mexico] 9 : 276–320.

Richards, J. V.

1911   "Dry Washing for Placer Gold in Sonora, Mexico." *Transactions*, American Institute of Mining Engineers 41 : 797–802.

Richardson Construction Company   « 179 »
    N.d.   *The Yaqui River Valley*. Los Angeles & New York.
Richardson, Rupert N.
    1933   *The Comanche Barrier to South Plains Settlement*. Glendale, Calif.: Arthur
                H. Clark Co.
Riley, Carroll L.
    1987   *The Frontier People: The Greater Southwest in the Protohistoric Period*. Al-
                buquerque: University of New Mexico Press.
Riley, Carroll L., and Basil C. Hedrick (eds.)
    1978   *Across the Chichimec Sea: Papers in Honor of J. Charles Kelley*. Carbondale:
                Southern Illinois University Press.
Ríos, Antonio
    1960   "El problema del salitre en los distritos de riego en México." *Ingeniería
                Hidráulica en México* 14(3) : 27–42.
Rippy, James F.
    1919   "Indians of the Southwest in the Diplomacy of the United States and
                Mexico, 1848–1853." *Hispanic American Historical Review* 2 : 363–396.
    1926   *The United States and Mexico*. New York: A. A. Knopf.
Robertson, William
    1777   *The History of America*. 2 vols. London: W. Strahan.
Roca, Paul M.
    1967   *Paths of the Padres through Sonora: An Illustrated Guide to Its Spanish
                Churches*. Tucson: Arizona Pioneers Historical Society.
Rosendal, Hans E.
    1963   "Mexican West Coast Tropical Cyclones, 1947–1961." *Weatherwise* 16 :
                226–230.
Rowan, John L.
    1962   "The Hermosillo Irrigation District: Problems and Prospects." M.A.
                thesis, University of California at Los Angeles.
Salazar, Luis
    1902   "Mexican Railroads and the Mining Industry." *Transactions*, American
                Institute of Mining Engineers 32 : 303–334.
Sanderson, Steven E.
    1981   *Agrarian Populism and the Mexican State: The Struggle for Land in Sonora*.
                Berkeley & Los Angeles: University of California Press.
Sandoval Godoy, Sergio Alfonso
    1988   "Los enlaces económicos y políticos de la Ford Motor Company en Her-
                mosillo." In J. C. Ramírez (ed.), *La nueva industrialización en Sonora: el
                caso de los sectores de alta tecnología*, pp. 133–203.
Santamaría, Francisco J.
    1959   *Diccionario de Mejicanismos*. Mexico City: Editorial Porrúa.
Sauer, Carl O.
    1935a   *Aboriginal Population of Northwestern Mexico*. Ibero-Americana, no. 10.
                Berkeley & Los Angeles: University of California Press.
    1935b   "A Spanish Expedition into the Arizona Apacheria." *Arizona Historical
                Review* 6 : 3–13.
    1941   "The Personality of Mexico." *Geographical Review* 31 : 355–364.

« 180 » Sauer, Carl O., and Donald Brand
1931   "Prehistoric Settlements of Sonora, with Special Reference to Cerros de Trincheras." *University of California Publications in Geography* 5(3): 67–148.

Schabel, Robert O.
1962   "Some Palms of Northwestern Mexico." *Principes* 6:5–9.

Shreve, Forest
1934   "Vegetation of the Northwestern Coast of Mexico." *Bulletin of the Torrey Botanical Club* 61:373–380.

Shreve, Forest, and I. L. Wiggins
1951   *Vegetation and Flora of the Sonoran Desert.* 2 vols. Publications of the Carnegie Institution of Washington, no. 591. Reprint 1964. Stanford, Calif.: Stanford University Press.

Shull, Dorothy B.
1968   "The History of the Presidios in Sonora and Arizona." M.A. thesis, University of Arizona, Tucson.

Signoret Vera, María Luisa
1965   "Aspecto geográfico del distrito de riego del Río Yaqui, Sonora." Professional thesis, Universidad Nacional Autónoma de México.

Siliceo, G. Manuel (ed.)
1857   *Memoria de la Secretaría de Fomento, Colonización, Industria y Comercio de la República Mexicana.* Mexico City.

Simpson, Lesley B.
1938   *Studies in the Administration of the Indians in New Spain. III, The Repartimiento System of Native Labor in New Spain and Guatemala.* Ibero-Americana no. 13. Berkeley & Los Angeles: University of California Press.

Smith, Mervin G.
1947   *Mexican Winter-Vegetable Export Industry.* Washington, D.C.: U.S. Dept. of Agriculture, Foreign Agriculture Report no. 21.

Sonora
1984   *Anuario estadístico.* 2 vols. Hermosillo: Gobierno del Estado de Sonora.
1990a   *Agenda estadística, 1989–1990.* Hermosillo: Gobierno del Estado de Sonora, Dirección de Información y Estadística.
1990b   *Cuaderno de Información para la planeación.* Aguascalientes: Instituto Nacional de Estadística.

Sorre, Max
1948   "La notion de genre de vie et sa valeur actuelle." *Annales de Géographie* 57:97–108, 193–204.
1962   "The Concept of Genre de Vie." In P. L. Wagner and M. W. Mikesell (eds.), *Readings in Cultural Geography*, pp. 399–415.

South, Robert B.
1990   "Transnational 'Maquiladora' Location." *Annals of the Association of American Geographers* 80: 549–570.

Southworth, John R.
1905   *Las minas de México* [in Spanish and English]. Liverpool: Blake & Mackenzie.

Spaulding, Alan
1974   "Agriculture in the Valle del Yaqui, Sonora, Mexico." M.A. thesis, California State University at Northridge.

Spicer, Edward H.
1962   *Cycles of Conquest: The Impact of Spain, Mexico, and the United States on the Indians of the Southwest, 1533–1960.* Tucson: University of Arizona Press.
1980   *Yaquis: A Cultural History.* Tucson: University of Arizona Press.

Standley, Paul C.
1920   *Trees and Shrubs of Mexico.* Pt. 1. U.S. National Museum, Contributions from the U.S. National Herbarium, vol. 23. Washington, D.C.: Smithsonian Institution.

Starkey, James H.
1980   "The Mexican Agricultural Economy: A U.S. Perspective." *Western Journal of Agricultural Economics* 5:209–218.

Steinbeck, John, and Edward F. Ricketts
1941   *Sea of Cortez.* New York: Viking Press.

Stewart, Norman R.
1955   "Alamos, Sonora: The Rise and Decline of a Mining Community in Northwest Mexico." M.A. thesis, University of California at Los Angeles.

Tamarón y Romeral, Pedro
1937   *Demonstración del vastísimo obispado de la Nueva Vizcaya* [1765]. Mexico City: Antigua Librería Robredo de J. Porrúa e hijos.

Thomson, Roberto
1989   *Pioneros de la Costa de Hermosillo (La Hacienda de Costa Rica 1844).* Hermosillo: Artes Gráficas y Editoriales Yescas.

Thrapp, Don L.
1967   *The Conquest of Apacheria.* Norman: University of Oklahoma Press.

Treutlein, Theodore E.
1939   "Economic Regime of the Jesuit Missions in Eighteenth Century Sonora." *Pacific Historical Review* 8:289–300.
1945   "Document: The Relation of Philipp Segesser (1737)." *Mid-America* 27:139–187, 257–260.
1965   *Joseph Och, S.J. Missionary in Sonora. The Travel Reports of Joseph Och, S.J., 1755–1767.* San Francisco: California Historical Society.

Treviño, Jorge E.
1982   "Programa de la fauna de acompañamiento del camarón en Guaymas: experiencia de investigación y desarrollo." *Gulf and Caribbean Fisheries Proceedings,* 1981, pp. 120–130.

Turnage, W. V., and T. D. Mallery
1941   *An Analysis of Rainfall in the Sonoran Desert and Adjacent Territory.* No. 529, pp. 1–45. Washington, D.C.: Carnegie Institution.

Ulloa, Pedro N.
1910   *El estado de Sonora y su situación económica.* Hermosillo: Imprenta del Gobierno.

« 182 »   Vargas Alcántara, Vicente

1959–1960   "Perforación de pozos profundos para explotación de agua subterránea."
Ingeniería Hidráulica en México, 13(2 & 3): 21–38, 77–96; 14(1): 71–103.

Velasco, José Francisco

1850   *Noticias estadísticas del estado de Sonora.* Mexico [City]: Imprenta de Ignacio Cumplido.

Vidal de la Blache, Paul

1903   *Tableau de la géographie de la France.* Vol. I, pt. 1. In Ernest Lavisse (ed.), *Histoire de la France illustrée.* Paris: Hachette.

1911   "Les Genres de vie dans la géographie humaine." *Annales de Géographie* 20: 193–213, 289–304.

1941   *La Personnalité géographique de la France.* Manchester: Manchester University Press.

Voss, Stuart F.

1982   *On the Periphery of Nineteenth Century Mexico. Sonora and Sinaloa, 1810–1877.* Tucson: University of Arizona Press.

Wagner, Philip L., and Marvin W. Mikesell

1962   *Readings in Cultural Geography.* Chicago & London: University of Chicago Press.

Waibel, Leo

1927   "Die nordwestlichen Küstenstaaten Mexikos." *Geographische Zeitschrift* 33: 561–576.

1928   "Die Inselberglandschaft von Arizona und Sonora." *Zeitschrift der Gesellschaft für Erdkunde zu Berlin, Sonderband,* pp. 68–91.

Waring, W. George

1897   "The Gold Fields of Altar, Mexico." *Engineering and Mining Journal* 63: 257–258.

Webber, Benjamin N.

1935   *Bajada Placers of the Arid Southwest.* American Institute of Mining and Metallurgical Engineers, Technical Publ. no. 588, Class I, Mining Geology, no. 51.

Wellhausen, E.

1976   "The Agriculture of Mexico." *Scientific American* 235(3): 128–150.

White, Steven S.

1948   "The Vegetation and Flora of the Rio Bavispe in Northeastern Sonora." *Lloydia* 11: 229–302.

Wilson, Eldred D.

1946   "Early Mining in Arizona." *The Kiva* 11(4): 39–47.

1952   *Arizona Gold Placers and Placering.* 5th rev. ed. Tucson: J. B. Tenney.

Wiseman, Frederick M.

1980   "The Edge of the Tropics: The Transition from Tropical to Subtropical Ecosystems in Sonora, Mexico." *Geoscience and Man* 21: 141–156.

Young, Richard H.

1982   "The Guaymas Shrimp Bycatch Program." *Gulf and Caribbean Fisheries Institute Proceedings (1981),* pp. 131–138.

Index

(Italicized page numbers refer to illustrations.)